OPTICAL MINERALOGY

The quality of the materials used in the manufacture of this book is governed by continued postwar shortages.

OPTICAL MINERALOGY

PUBLISHED FORMERLY UNDER THE TITLE
THIN-SECTION MINERALOGY

BY

AUSTIN F. ROGERS, Ph.D.
Professor of Mineralogy and Petrography
Stanford University

AND

PAUL F. KERR, Ph.D.
Professor of Mineralogy, Columbia University

SECOND EDITION
FOURTH IMPRESSION

McGRAW-HILL BOOK COMPANY, Inc.
NEW YORK AND LONDON
1942

THE MAPLE PRESS COMPANY, YORK, PA.

TO

THE MEMORY OF

LEA McILVAINE LUQUER

1864–1930

PREFACE TO THE SECOND EDITION

Seven years have elapsed since the first edition of this text appeared. During this time the writers have had an opportunity to discuss with various coworkers in this field the treatment adopted in the first edition, and as a result numerous suggestions kindly offered have been taken into consideration in this revision. In particular, we wish to acknowledge the cooperation of Dr. M. N. Short of the University of Arizona, Dr. Ian Campbell of the California Institute of Technology, Dr. Howard A. Coombs of the University of Washington, Dr. A. O. Woodford of Pomona College, Dr. J. J. Runner of the State University of Iowa, Dr. George Tunell and Dr. F. E. Wright of the Geophysical Laboratory, Dr. Aaron C. Waters of Stanford University, Dr. S. J. Shand of Columbia University, Dr. W. T. Schaller, and Dr. Clarence Ross of the U. S. Geological Survey.

The general form of treatment followed in the first edition has been retained, but many explanations of optical properties have been rewritten, and a considerable number of new diagrams have been added. Descriptions of a number of additional minerals have been added, and the descriptions of a number of groups have been rewritten. An effort has been made, however, to avoid lengthening the text to any considerable extent.

The first edition was restricted to the utilization of optical properties in the identification of minerals in thin sections. Experience has shown, however, that the text has found application in other forms of mineral identification with the microscope. In recognition of this situation the title has been changed, and certain portions of the text have been reorganized to extend its field of usefulness, although it still applies primarily to mineral identification in thin sections.

A. F. R.,
P. F. K.

STANFORD UNIVERSITY,
COLUMBIA UNIVERSITY,
  January, 1942.

PREFACE TO THE FIRST EDITION

The identification of many of the more common minerals encountered in thin sections of rocks may be accomplished by using simplified methods of optical mineralogy. It is not necessary to require the student to spend a large amount of time in studying the great volume of theoretical information necessary to acquire an advanced knowledge of the optics of crystals. If a more advanced knowledge of the subject is desired, however, practice in the optical identification of the common minerals is the first step in approaching the advanced phases of the subject.

This text draws upon material used in introductory courses of mineral optics both at Columbia University and at Stanford University. The treatment is intended to develop optical mineralogy for the beginning student, whether he happens to be interested primarily in the field of mineralogy, petrography, geology, chemistry, physics, or engineering. The principles outlined apply considerably beyond the field of mineralogy, although the application is restricted to minerals.

The discussion of the subject has been made as nearly non-mathematical as possible for the sake of simplicity. A large number of illustrations have been used in proportion to the text since many features are best explained by diagrams. Data concerning the identification of minerals in fragments have been included to be of assistance in determining minerals not easily identified in thin section alone. Charts and tables are used to increase the speed of mineral determination by outlining a systematic procedure.

The preparation of the manuscript has been greatly aided by the friendly criticism of a number of coworkers in the field of optical mineralogy or petrography. Dr. George Tunell of the Geophysical Laboratory has read the chapters concerning optical theory and offered numerous suggestions. He has kindly discussed portions of the manuscript with several of his colleagues at the Geophysical Laboratory and has commented upon the results of such discussions. Dr. Clarence S. Ross of the U. S.

Geological Survey examined the text prior to publication. Dr. Waldemar T. Schaller of the U. S. Geological Survey, Professors R. J. Colony and William M. Agar of Columbia University, Dr. J. W. Greig of the Geophysical Laboratory, Professor Aaron C. Waters of Stanford University, and Professor L. E. Spock of New York University have also suggested a number of features. All have contributed most generously of their time and experience, for which we wish to express our appreciation. We are indebted to Mr. Paul H. Bird, graduate student of Columbia University, for advice in describing the technique of making thin sections. The line drawings of Part II have been made by Mr. Rudolph G. Sohlberg. In the proof-reading of Part II we gratefully acknowledge the assistance of Miss Genevieve Rogers. The various optical companies mentioned in Chap. II have generously cooperated by furnishing illustrations of optical equipment.

In writing the text the description of the optical properties of mineral species (Part II) have been prepared by the senior author. The junior author has contributed the discussion included in Part I and also the determinative tables.

A. F. R.,
P. F. K.

Stanford University,
Columbia University,
    October, 1933.

CONTENTS

PART I

MINERAL OPTICS

CHAPTER I

CHAPTER II

CHAPTER III

CHAPTER IV

CHAPTER V

PART II
DESCRIPTIONS OF INDIVIDUAL MINERALS

TABLE OF ABBREVIATIONS

Symbols for Indices of Refraction in General Use

Mineral type to which index symbol applies	Symbols used in this text	Symbols used by Dana, Johannsen, Larsen, and Berman	Symbols used by Winchell
Isotropic...................	n	n	N
Uniaxial			
Extraordinary ray..........	n_ϵ	ϵ	Ne
Ordinary ray..............	n_ω	ω	No
Biaxial			
Least value...............	n_α	α	Np
Intermediate value.........	n_β	β	Nm
Greatest value............	$n\gamma$	γ	Ng

n = index of refraction.

n_α(alpha) = the index of the fast ray in biaxial minerals. The least index of refraction.

n_β(beta) = the index of the ray at right angles to n_α and n_γ.

n_γ(gamma) = the index of the slow ray in biaxial minerals. The greatest index of refraction.

n_ϵ(epsilon) = the maximum (in positive) and the minimum (in negative) index of refraction of the extraordinary ray in uniaxial minerals.

n_ω(omega) = the index of refraction of the ordinary ray in uniaxial minerals. If $n_\omega < n_\epsilon$, the mineral is positive. If $n_\omega > n_\epsilon$, the mineral is negative. n_ω is constant in a given uniaxial mineral, whereas the index of the extraordinary ray varies from n_ω to n_ϵ.

n_1 and n_2 = the lesser and greater indices of refraction of the two rays in any crystal section at random orientation.

X = the axis of greatest ease of vibration. Light vibrating parallel to X travels with maximum velocity (also indicated by α).

Z = the axis of least ease of vibration. Light vibrating parallel to Z travels with minimum velocity (also indicated by γ).

Y = the intermediate axis at right angles to the plane of X and Z (also indicated by β).

ϵ = the axis of vibration of the extraordinary ray.

ω = the axis of vibration of the ordinary ray in a plane at right angles to ϵ.

r = the dispersion for red.

v = the dispersion for violet.

2V = the axial angle within the mineral.

2E = the axial angle observed in air.

Bx_a = acute bisectrix.

Bx_o = obtuse bisectrix.

Ax. pl. = the plane of the optic axes.

μ = micron, thousandth of a millimeter (0.001 mm.).

$m\mu$ = millimicron, millionth of a millimeter (0.000001 mm.).

AU = angstrom unit, tenth of a millimicron (0.0000001 mm.).

Δ = retardation in $m\mu$ (millimicrons).

t = thickness of a thin section.   Usually given in hundredths of a milli-
    meter (0.01 mm.).

a, b, and c = the crystallographic axes.

$\angle \alpha$, β, γ = angles between the crystallographic axes.

$(n_\gamma - n_\alpha)$ = double refraction for biaxial minerals.

$(n_\omega - n_\epsilon)$; $(n_\epsilon - n_\omega)$ = double refraction for uniaxial minerals.

H_1 = the slow ray of the Berek compensator.

H_2 = the fast ray of the Berek compensator.

e = the extraordinary ray.

o = the ordinary ray.

Length-fast (or negative elongation) = elongation parallel to the vibration
    direction of the fast ray.

Length-slow (or positive elongation) = elongation parallel to the vibration
    direction of the slow ray.

ca = circa (about).

PART I
MINERAL OPTICS

CHAPTER I

THE PREPARATION OF THIN SECTIONS OF MINERALS AND ROCKS

Although students of mineralogy seldom prepare their own thin sections, every worker should know at least in a general way how thin sections are made. A student with natural dexterity may learn to make thin sections approaching those ground by professional section makers.[1] The speed may not equal that of the professional, but with a little care the section made may be just as good. A brief discussion of the methods used in section making will be given for the benefit of the beginner.

The technique employed depends largely upon the nature of the material. Grinding sections of ordinary igneous, sedimentary, and metamorphic rocks is a routine process. Friable or fractured rocks and unconsolidated materials, however, require special precautions not necessary in the case of compact specimens.

The various stages in the preparation of an ordinary rock section are illustrated in Fig. 1. The first stage represents the choice of material. A specimen suitable for effective study with the microscope is selected. Such a specimen may contain fine-grained minerals not easily studied by the unaided eye, structures that yield readily to examination with the microscope, or any one of those numerous minor features so effectively revealed by microscopic examination.

A chip of suitable size should be sawed or broken from the specimen. An ideal chip is flat and almost square, about $\frac{1}{8}$ in. thick and 1 in. square. A chip with flat surfaces can be prepared much better by sawing with a rock saw than by break-

---

[1] The following list is furnished for the convenience of readers who wish the names of technicians making thin sections:

George Rév, 1180 Amsterdam Ave., New York.
Alexander Tihonravov, 27 Van Buren, Los Altos, Calif.
W. Harold Tomlinson, Swarthmore, Pa.
Rudolph von Huene, 865 N. Mentor Ave., Pasadena, Calif.

3

ing.   Either a diamond saw or a carborundum saw is satisfactory for this work.   Ordinary saws consist of thin wheels of soft metal, impregnated on the margins with abrasive powder.   The impregnated carborundum saw is ordinarily used since it is much less expensive and is entirely satisfactory for routine work.

Figure 2 illustrates an ordinary type of mineralogical saw. It may be a hard-rolled copper disc that operates over a trough

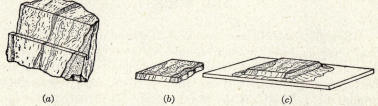

    (a)   (b)   (c)

Fig. 1.—(a) A rock selected for a thin section and sawed to obtain a chip. (b) A sawed chip of proper dimensions for a thin section (bottom surface polished). (c) The chip mounted on a glass slide with Canada balsam ready for the first stage of grinding.

filled with carborundum and sludge.   The edge of the rotating disc picks up carborundum and rubs it against the specimen. If a proper mixture of mud, carborundum, and water is maintained, it is possible to grind a narrow channel completely through an inch of solid quartz in 5 minutes.

When the chip is ready, a smooth surface is polished on one side by utilizing successively 100, FFF, and 600 carborundum

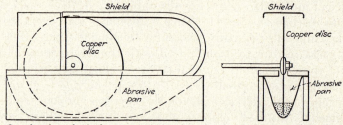

Fig. 2.—A mineralogical saw consisting of a hard-rolled copper disc used with carborundum.

and finishing with 302½ American Optical Company's emery. In case the rock is fairly soft, the first grinding with 100 carborundum is omitted.   The 100 carborundum is too coarse and tends to destroy soft material.

Mechanical grinding of thin sections usually takes place on flat laps constructed of suitable metal and faced to a plane flat surface. The laps should be at least 12 in. in diameter and should rotate at a speed of about 600 r.p.m. Bearings should be shielded against abrasive powder, and drive shafts should be arranged to transmit the power to the lap by friction if possible. The lap for fine grinding should be made of copper or brass and grooved.

Great precautions concerning cleanliness are necessary throughout the entire process of preparing a thin section. A single grain of coarse grit rubbed against the slide at the wrong stage of the process will often destroy a thin section.

The smoothly ground but unpolished surface is cleaned and dried. It is then mounted on a glass object slide, employing Canada balsam as a cementing material. The balsam should be cooked slowly until a chilled drop will snap from the end of a needle when pressed against the thumbnail.[1] It should not be overcooked since it then becomes too brittle and may even turn brown. While the balsam is still warm and liquid, the flat surface of the warmed chip is placed upon a slide having a cooked smear of Canada balsam. On cooling, the chip will be firmly cemented to the glass slide. The bond should be an even layer of balsam unbroken by air bubbles. If air bubbles are present, the chip should be warmed, removed, and remounted. The problem of properly cooking Canada balsam may be solved by using a plate with an interior chamber filled with glycerin to

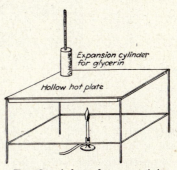

Fig. 3.—A hot plate containing a glycerin chamber and thermometer well for proper control in cooking Canada balsam. (*Developed by Paul H. Bird.*)

distribute the heat evenly, together with an expansion cylinder containing a thermometer for temperature control (Fig. 3).

The exposed side of the mounted chip is ground with medium and fine carborundum and alundum. Alundum is utilized for the final grinding when the chip has reached a thickness of about 0.1 mm. The specimen is usually ground on a rotating lap and finished by hand on a smooth glass plate. A fine alundum or

[1] Balsam is first cooked about 2 min. at 160° C. It may then be warmed at 120° for mounting and heated again at 100° to attach the cover glass.

emery paste is used on the glass plate. The final grinding demands manual dexterity. The thin slice should be kept uniform in thickness during grinding and the grinding continued until a thickness of about 0.03 mm. is attained. The thickness of the slide is controlled through the final stage by microscopic observation of the interference colors given by some known mineral in the section when covered with a film of water. Quartz is frequently present, in which case the interference colors due to this mineral should be almost entirely white or gray.

Small holders are useful for grinding the mounted chips. Such holders keep the opposite surfaces of the slide parallel and facilitate the process of fine grinding. Holders also permit the operator to grind several sections at the same time until the final stage is reached. Three holders form a useful combination.

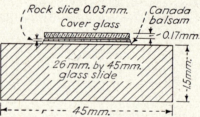

FIG. 4.—A cross section of the mounted rock slice (vertical scale exaggerated).

One holding six slides may be used to grind the chip to a thickness of about 0.5 mm.; the second, arranged for two slides, to carry the grinding to 0.1 mm.; and the third, holding a single slide, to complete the section. In case a holder is not available, a small cork may be cemented to the back of the slide with balsam.

When the section is ground to the proper thickness, it is washed free from grinding powder and dried. Fresh balsam is then smeared over the surface of the slice, cooked, and a cover glass (thickness about 0.17 mm. or less) placed over it. The preparation is then cooled, and excess balsam around the edge of the cover glass is dissolved with xylol, followed by a wash with kerosene. The thin section is now covered and ready for use (Fig. 4).

In case the slice is composed of substantial and compact material, it may be transferred from the glass slide on which it has been ground to another slide free from scratches. To effect the transfer, the slide holding the slice is smeared with balsam and heated on a hot plate as in covering. A clean slide smeared with balsam is placed on the hot plate next to the ground slide containing the chip, and the balsam cooked at the same time. When both smears of balsam are cooked, the slice is worked free from the ground slide with the aid of a toothpick and floated to the clean slide. It is then covered with a cover glass in the usual way.

Poorly consolidated materials require special treatment. Dr. C. S. Ross has developed the technique of making thin sections of friable materials in the petrographic laboratory of the U. S. Geological Survey. His methods apply to such substances as incoherent sand, tuff, soil, plastic clay, shale.

Two main problems are involved. The first is impregnating with some substance that will act as a binder; the second is to use grinding media that will not react with the material of the thin section and cause its disintegration. Kollolith,[1] balsam, or bakelite may be used as binders. Petroleum oil, alcohol, ether, acetone, glycol, or water may be used for grinding media, depending upon the character of the material. Carborundum paper, carborundum, emery, or alundum serves as an abrasive, and the grinding may be quickly done by hand on glass plates.

Thin "peeled" films removed from the smooth surfaces of clays may be mounted in balsam on glass slides and examined as ordinary thin sections. The mounted films are for the most part exceedingly thin, although coarse particles may be plucked from the clay mass and will protrude from the balance of the film.

A smooth surface is carefully prepared by polishing on dry ground-glass plates. While still moist, the surface is covered with amyl acetate and pyroxylin. After the preparation is dried for from 5 to 6 hr., the dry film formed by the pyroxylin is peeled from the clay surface with a knife blade. The film is then mounted with Canada balsam on a glass slide and covered with a cover glass.

Suggested References

HOLMES, A.: "Petrographic Methods and Calculations," pp. 231–249, Thomas Murby and Co., London, 1930.

JOHANNSEN, A.: "Manual of Petrographic Methods," 2d ed., pp. 572–604, McGraw-Hill Book Company, Inc., New York, 1918. (With bibliography.)

[1] Manufactured by Voigt and Hochgesang, Göttingen, Germany.

KEYES, MARY G.: Making Thin Sections of Rocks, *Am. Jour. Sci.*, 5th ser., vol. 10, pp. 538–550, 1925.

Ross, C. S.: Methods of Preparation of Sedimentary Materials for Study, *Econ. Geol.*, vol. 21, pp. 454–468, 1926. See also *Am. Jour. Sci.*, 5th ser., vol. 7, pp. 483–485, 1924.

WEATHERHEAD, A. V.: A New Method for the Preparation of Thin Sections of Clays, *Mineralog. Mag.*, vol. 25, pp. 529–533, 1940.

WEYMOUTH, A. ALLEN: Simple Methods for Making Thin Sections, *Econ. Geol.*, vol. 23, pp. 323–330, 1928.

CHAPTER II

THE POLARIZING MICROSCOPE

General Discussion.—The polarizing, or the petrographic, microscope, as it may be called, is used to the exclusion of other models in the study of thin sections of minerals and rocks. The lens system is similar to the lens system of the usual modern compound microscope. The instrument, however, contains several additional features that greatly increase its range of usefulness. The most distinctive are the polarizing and analyzing prisms and several accessories such as the Bertrand lens, mica plate, gypsum plate, and quartz wedge.

The names applied to the various parts of a polarizing microscope are given in Fig. 5. The microscope illustrated may be considered either as a student model or as a model designed for routine forms of microscopic work.[1] It is only in cases of advanced research that a more elaborate instrument is required.

The polarizing microscope is adapted for ordinary examination of minerals in plane-polarized light, observation between crossed nicols, and conoscopic study. The lower polarizing prism is left in place beneath the condenser, and the upper prism is moved to one side for ordinary inspection. A full range of magnifications becomes possible with this arrangement. When the microscope is used for examination of specimens between crossed nicols,[2] both prisms are inserted in the path of light. The path of light through the microscope with the latter set-up is shown in the sectional view (Fig. 7). Observation between crossed nicols is

---

[1] Polarizing microscopes in common use are manufactured by:

Bausch and Lomb Optical Co., Rochester, N.Y.
Carl Zeiss, Inc., 485 Fifth Ave., New York, N.Y.
E. Leitz, Inc., 730 Fifth Ave., New York, N.Y.
Spencer Lens Co., Buffalo, N.Y.

[2] The word *nicol* is frequently used in a broad sense in referring to either the polarizing or the analyzing prisms. The prism used, however, may not be, strictly speaking, a Nicol prism.

possible over the same range of magnifications as that employed for ordinary examination. A simple lens is often inserted above the analyzing prism to correct for the slight change in magnification due to the prism.

Conoscopic study consists of examination between crossed nicols in convergent polarized light with a Bertrand lens. The sequence of units employed for the conoscopic set-up is illustrated

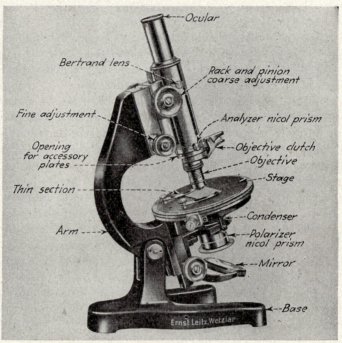

Fig. 5.—A student model petrographic microscope with parts marked. (*E. Leitz, Inc.*)

in Fig. 8. The 16-mm. objective illustrated in the figure is serviceable for crystals of large areas. Ordinarily, either 8- or 4-mm. objectives are used in obtaining interference figures from small crystals.

Parts of the Microscope. *Oculars.*—Oculars used in modern petrographic microscopes are ordinarily of the Huygenian type or a simple modification. In combination with 40- and 16-mm. objectives or other objectives in the same range of magnification, Huygenian oculars are employed. Where combinations giving higher magnifications are desired, the ocular is similar to the Huygenian ocular but contains a specially corrected eye-lens arrangement giving a flat field. This adjustment is particularly important for photomicrography.

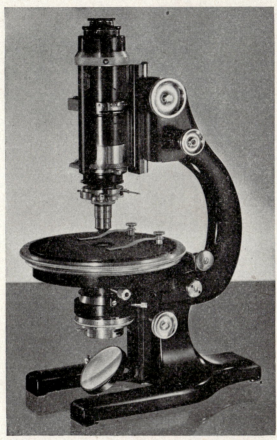

Fig. 6.—A student model polarizing microscope.   (*Spencer Lens Company.*)

In the Huygenian ocular the stop is located between the two lenses. An ocular of this type is frequently called a *negative ocular*. The Ramsden ocular with the stop below the two lenses is in contrast described as a *positive ocular*. The arrangement of the stops in the two types is shown in Fig. 9: *a* represents the Huygenian ocular and *b*, the Ramsden type.

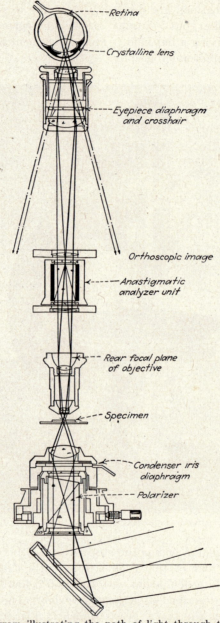

Fig. 7.—A diagram illustrating the path of light through the microscope for ordinary observation with crossed nicols.   (*Spencer Lens Company.*)

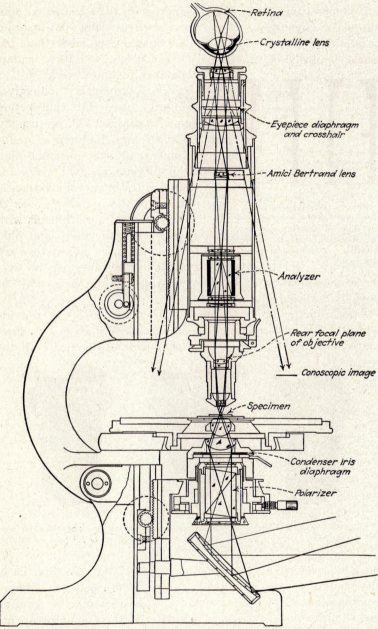

Retina

Crystalline lens

Eyepiece diaphragm
and crosshair

Amici Bertrand lens

Analyzer

Rear focal plane
of objective

Conoscopic image

Specimen

Condenser iris
diaphragm

Polarizer

FIG. 8.—A sectional view of the polarizing microscope as a conoscope for the study of interference figures and also for high magnifications. (*Spencer Lens Company.*)

Compensating oculars are constructed to accompany apochromatic objectives. Some manufacturers claim that in order to secure the best results, oculars magnifying more than ten times should be of this type. Ordinary 5 × and 10 × oculars are satisfactory for most work.

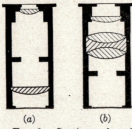

Objectives.—Achromatic objectives are ordinarily used for thin-section or fragment studies. Those usually supplied by the manufacturers as standard equipment, 40-, 32-, 16- and 4-mm. achromatic objectives, will serve for most studies. In the case of achromatic objectives correction of aberrations of the image becomes more difficult with high eyepiece magnification, and only the best achromatic objectives will give satisfactory results with an eyepiece magnification of 12 × or greater. Views of several cut objectives appear in Fig. 10.

FIG. 9.—Sections of positive and negative oculars. (a) The Huygenian ocular (a negative ocular). (b) The Ramsden ocular (a positive ocular).

Apochromatic objectives have been constructed to provide additional color correction beyond that usually given by achromatic objectives. In this type of objective practically all the

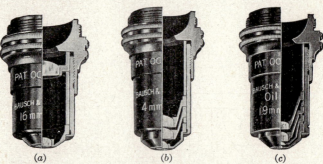

FIG. 10.—Sectional views of objectives. (a) Achromatic objective 16 mm. (b) Achromatic objective 4 mm. (c) Apochromatic objective 1.9 mm. oil immersion. (*Bausch and Lomb Optical Company.*)

images produced by the different colors of the spectrum lie in the same plane and are equally sharp. The lenses are made of combinations of fluorite and glass. The difficulty of securing good fluorite and the practical difficulties in their manufacture are considerable; consequently the cost is greater than the cost of ordinary achromatic objectives. However, these objectives are seldom necessary for microscopic study of minerals.

The principal features of an objective that are of interest to the student are the initial magnification, the numerical aperture, the focal length, and the working distance.

The focal length may be employed in determining the approximate initial magnification of an objective. The optical tube length divided by the focal length equals the initial magnification. Several manufacturers stamp the initial magnification for a standard mechanical tube length[1] on the objective. This figure multiplied by the power of the eyepiece gives the magnification for a standard tube length. This should be corrected, however, when the analyzing prism is inserted (unless the prism mount contains a correcting lens). Corrections can be determined by using stage and eyepiece micrometers.

The working distance is the distance between the objective and the top of the cover glass of the microscope slide when the objective is in focus.

The numerical aperture (N. A.) of an objective is a measure of the largest cone of light that it covers from an object point at the principal focus. N. A. equals $n \sin \mu$, where n is the index of refraction of the medium between the object under examination and the objective[2] and μ is one-half the angle of the cone of light entering the lens. The numerical aperture furnishes a criterion of the quality of an objective. Other things being equal, at any magnification, the intensity of the image is proportional to N. A.; the resolving power is directly proportional to N. A.; the depth of focus is inversely proportional to N. A. In two objectives having the same focal distance and therefore the same magnification, the one with the greater N. A. will take a larger cone of light from the object and will yield a brighter image. In general, with ordinary lighting, the limit of useful magnification for an average observer is between 500 times the N. A. and 1000 times the N. A.

Oil-immersion objectives are used for high magnifications where a high degree of resolving power and correction are required.

---

[1] Bausch and Lomb Optical Co. and Spencer Lens Co. = 160 mm. Leitz = 170 mm. Zeiss = 170 mm.

[2] Air ($n = 1$) in the case of a dry objective and specially prepared cedar oil ($n = 1.515$) in an oil-immersion objective.

The oil should agree in both dispersive power and index of refraction with the front lens of the objective. The effect of oil immersion on the cone of light entering the front lens of an oil-immersion objective is shown in Fig. 11. A considerable advantage is also gained by placing a drop of oil between the auxiliary condenser lens and the microscope slide. The working distance of an oil-immersion objective is very short; the lenses are difficult to manufacture and are consequently expensive. A good oil-immersion objective, however, gives a beautiful field with high magnification. The objective should be handled carefully, especially in focusing. After use the oil should be removed by the use of lens paper moistened with xylol or benzine.

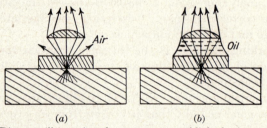

(a)   (b)

Fig. 11.—Diagram illustrating the convergence of light by means of cedar oil placed in front of the lens of an oil-immersion objective. (a) Air alone without cedar oil; (b) with cedar oil.

Microscope Accessories.—The accessories provided with the microscope generally include a quartz wedge, gypsum plate, and mica plate. These are marked with arrows indicating the fast- and slow-ray vibration directions and are mounted in frames to fit the opening in the tube of the microscope between the objective and the analyzer.

The quartz wedge is ground to produce interference colors from the beginning of the first to the end of the third or fourth order. It is marked and mounted as shown in Fig. 12.

The mica plate and gypsum plate (German = *Glimmer* and *Gips*) together with a centering pin are illustrated in Fig. 13. *N* is the slow-ray direction in the mica and gypsum plates (Leitz microscope).

Analyzer.—The prism mounted in the tube of the microscope above the objective is known as the *analyzer*. It is usually carried on a sliding mount so that it may be inserted or withdrawn from the optical axis at will. The plane of vibration is usually either perpendicular or horizontal in the field of view. More elaborate microscopes are fitted with a means for rotating the analyzer through 90°.

Polarizer.—The prism mounted in the substage system below the condenser is known as the *polarizer*. It is arranged for

FIG. 12.—The quartz wedge mounted on a glass plate and in a metal frame. (*Spencer Lens Company.*)

any adjustment through 360° but is usually kept adjusted to a plane at right angles to the plane of the analyzer. The position of the polarizer in its mounting is shown in Fig. 15.

Bertrand Lens.—This lens is inserted in the tube of the microscope between the ocular and the analyzer. It serves to bring

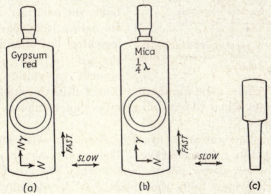

FIG. 13.—The gypsum plate (*a*), mica plate (*b*), and a centering pin (*c*). (*E. Leitz, Inc.*)

the image of an interference figure into the focal plane of the ocular.

Interference figures may be observed without the Bertrand lens if the ocular is removed. For best results, a Bertrand lens with a focusing diaphragm and an auxiliary magnifier to fit over the eyepiece is used.

Condenser.—Three components may be present in a condenser system of the type selected for illustration. In ordinary examination with low-power objectives a lens component with an illuminating aperture of about 0.22 is used. In working with high power or in obtaining interference figures, another condenser on a movable mounting swings across the axis (Fig. 14). This suffices for all objectives of N. A. up to 1.0. In the case of higher numerical apertures a special lens is inserted in place of the condenser in the movable mounting. This is more effective if used with oil immersion.

The arrangement of the condenser, together with the various adjustments for the polarizer, is shown in Fig. 15.

Iris Diaphragm.—The iris diaphragm is attached to the lower side of the tube that holds the polarizer. It serves to reduce the cone of light, lessening the illumination of the field of view, and causes objects to stand out with increased relief. The diaphragm is useful in the application of various tests when determining indices of refraction with the microscope.

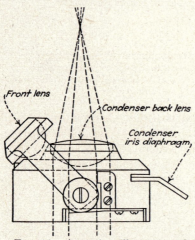

Fig. 14.—A section illustrating the construction of the condenser system. (*Spencer Lens Company.*)

Mirror.—The mirror is usually reversible, with one surface plane and the other concave. The plane mirror surface is suitable for low-power microscopic work. The concave mirror converges the light upon the object. It is especially useful in high-power examination. It should also be used for low power when the illuminator produces a convergent beam.

Fine Adjustment.—It is advantageous to have the fine adjustment graduated so as to permit the measurement of the displacement of the tube to within 2.5 μ (thousandths of a millimeter). The adjustment is used both for measuring depth and for focusing on objects at high magnifications. The relationship between a coarse and fine adjustment and the detail of the fine adjustment for one type of microscope are illustrated in Fig. 16.

Berek Compensator.—The compensator is designed to fit the tube slit above the objective in the same opening used for the gypsum and the mica plates. It is employed in the determination of the order of interference colors between crossed nicols.

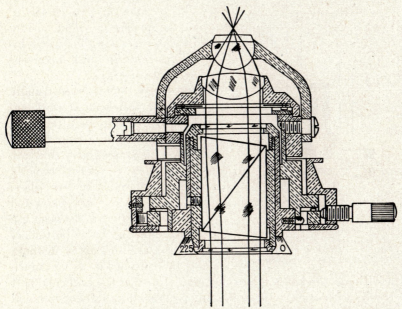

Fig. 15.—The arrangement of the polarizer below the condenser. (*Spencer Lens Company.*)

The plate is inserted with the vibration directions opposed to those of the mineral being examined. The compensator is adjusted until the color of the mineral is neutralized (becomes gray). The amount of adjustment of the compensator necessary to bring this about is a measure of the retardation. A view of the Berek compensator appears in Fig. 17.

Object Slide.—Various lengths and widths of object slides may be used, but the thickness is of greater importance. Immersion condensers are made to work to best advantage with slides from 0.9 to 1.0 mm. thick. Thus slides intended for study at high magnifications should conform to this thickness if the most satisfactory results are to be secured.

Slides 26 mm. wide by 45 mm. long are generally used for mounting thin sections of minerals and rocks. Such slides fit easily on the rotating stage of the polarizing microscope yet are large enough to contain a good-sized slice and also a label of suitable dimensions. Long slides usually employed in biological investigations may be quite inconvenient on a rotating stage.

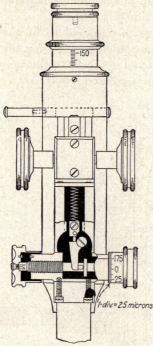

FIG. 16.—The fine adjustment. (*Spencer Lens Company.*)

Cover Glass.—Objectives usually employed for thin-section work are corrected by the manufacturers for a cover-glass thickness of from 0.15 to 0.17 mm. It is assumed that the top of the slice is pressed directly against the bottom of the cover glass. In case the slide is poorly mounted and a space intervenes between the top of the slice and the bottom of the cover glass the extra distance should be considered as so much additional thickness of cover glass. In order to obtain the best results with objectives, cover glasses of standard thickness should be employed.

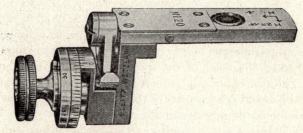

FIG. 17.—The Berek compensator.   (*E. Leitz, Inc.*)

Precautions to Be Observed in the Use of the Microscope.—Even under the best conditions microscope work produces a certain amount of strain upon the eyes. It is essential, therefore, to employ the best possible conditions of work in order to reduce such strain to a minimum.

The student should assume an erect but not too rigid position. Such a position with the microscope tube inclined allows him to work with maximum comfort.

Both eyes should be kept open while looking through the instrument. If it is difficult to do this at first, a shield should be placed over the eye not in use. A binocular eye guard is illustrated in Fig. 18. A spring-clamp ring for attachment to the

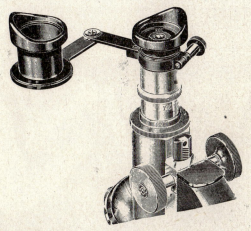

FIG. 18.—A binocular eye guard for the unused eye in microscope work.

microscope tube carries a lateral-jointed arm with a dummy eyepiece for covering the unused eye. It is also a good plan to learn to observe equally well with either eye and not to develop the so-called *microscope eye*.

In laboratories requiring a large amount of routine microscopic work an attachment known as the *euscope* has been devised for projecting the image of the field on a small shielded viewing screen. The observer is seated directly in front of the screen and looks forward into a viewing box with the image on a screen at the opposite end, as shown in Fig. 19.

The euscope has several applications for routine work. A ground glass may be inserted on the front of the instrument and used for microprojection. Grating lines may be ruled on the ground glass corresponding to the divisions of a stage micrometer, in which case the euscope will serve for microscopic measurement of grain size.   The instrument is also adapted to hold a camera for taking photomicrographs of thin sections.

Care of the Instrument.—A polarizing microscope is an expensive piece of equipment.   Properly used, it should last a lifetime. If not handled carefully, it may become useless with very little real service.   Most of the precautions to be observed in the use of the instrument are such as should be applied to any piece of fine apparatus.   A few, however, are of special nature and should be definitely mentioned.

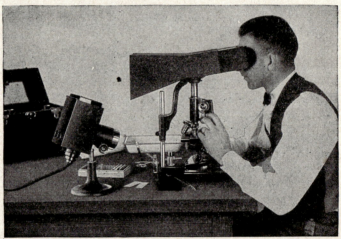

FIG. 19.—The euscope developed originally for biological work but found to be useful for mineralogical work as well.   (*Bausch and Lomb Optical Company.*)

Fine-textured lens paper, or, still better, a camel's-hair brush, should be used for cleaning all optical parts.   This applies to the ocular, the objectives, the substage system, the mirror, and the two nicols.

Objectives should be brought into focus by moving the tube of the microscope upward rather than downward.   Possibility of contact between the lower lens of the objective and the thin section is thus avoided.   High-power or oil-immersion objectives should be cleaned with lens paper and xylol or benzine (not alcohol).

Chemicals should not be used on the stage unless special precautions are taken to protect the objective.   Objectives may be protected by the use of cover glasses fastened to the lower lens. Occasionally an old objective is reserved for chemical work alone.

Magnification.—The microscope is primarily an instrument for magnification. It is worth while, therefore, to form at the outset an idea of the enlargement of the field of view with the various lens systems available. The following table outlines the various magnifications at the eye for different combinations of objectives with an equivalent focus of 40, 32, 16, 8, 4, and 2 mm. (oil immersion) and also oculars magnifying five, ten, and fifteen times, respectively.

MAGNIFICATIONS[1]

Type of objective	Equivalent focus, millimeters	Magnification number	Magnifications with oculars			Working distance, millimeters	N. A.
			5×	10×	15×		
Achromatic........	40	3.2	16	32	48	34.5	0.12
Achromatic........	32	4.3	22	43	65	27.0	0.15
Achromatic........	16	10	50	100	150	5.8	0.25
Apochromatic......	8	23	115	230	345	0.85	0.65
Apochromatic......	4	46	230	460	690	0.20	0.95
Apochromatic (oil immersion)......	2	92	460	920	1380	0.11	1.32

Tube length: 170 mm.
Image distance: 250 mm.
[1] After Leitz.

There are certain definite limits to the resolving power of the microscope, even with the best lens systems available. As long as the increase in magnification results in better vision of an object and more definite separation of detail, the magnification may be said to be "useful." When the object merely becomes larger without any increase in resolving power, the magnification is "empty." So-called, *empty* magnifications of great magnitude are possible.

For practical purposes the upper limit of "useful" magnification with the polarizing microscope is about 1800:1.[1] Larger

[1] An oil-immersion objective (Carl Zeiss) primary magnification 120, N. A. 1.3, free working distance 0.08 mm., in combination with a 15 × ocular, should yield a magnification ratio of 1800:1.

magnifications, as usually reported, are the result of some form of projection or special equipment in which the exact limits of useful magnification are not clearly known. The most ordinary form of projection is the enlargement employed in taking photomicrographs. Photomicrographs taken with a camera having a long bellows may increase the magnification ratio given by the microscope several times. Thus magnification ratios of 3000:1, 4000:1, or even considerably higher may be obtained. Such increase in magnification above the magnification of the microscope is essentially enlargement and does not result in increase in resolution. From the standpoint of increase in resolution or detail, it is "empty" magnification. Enlarged photomicrographs of this type, however, may have value for purposes of demonstration.

The limit of resolution for green light with a lens of N. A. 1.40 is said to be approximately 0.18 μ. This might be described as the distance apart of two object points in the field of view of the microscope whose disc images would just touch as projected to the eye. It has been shown mathematically that the limit of resolution equals the wave length divided by twice the numerical aperture. From this relationship it is possible to compute the number of lines per inch that can be separated by different numerical apertures. Several may be given as follows for blue light wave length 486:

N. A.	Lines per Inch Separated
1.30	136,000
0.85	89,000
0.65	68,000
0.30	31,000

An accurate check of the magnification of the field of view in the microscope can be obtained by using a stage micrometer (Fig. 20). The stage micrometer is a glass slide carefully ruled into hundredths of a millimeter. It not only serves as a comparison object for determining the magnification of the microscope but also may be used to give the magnification of micro drawings, of micro projections, and of photomicrographs.

Micrometer eyepieces are also utilized when the dimensions of particular objects in the field of view are desired (Fig. 21*a*). Such eyepieces are useful in determining the axial angle of interference figures with the microscope. The eyepieces should be calibrated with the aid of the stage micrometers for various objectives. The dimensions represented by the divisions in the micrometer ocular (Fig. 21b) as observed at the eye are governed by relations between the objective, the eyepiece, the tube length, and by the presence or absence in the optical train of the analyzer.

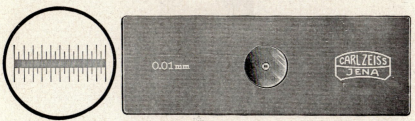

FIG. 20.—The stage micrometer. (*Carl Zeiss, Inc.*)

Micrometer eyepieces of the grating type (Fig. 21c) are employed to measure the areas of grains or fragments in the microscope field. These are also calibrated for different lens combinations with a stage micrometer.

Illumination.—At ordinary magnifications a good north light with a broad, clear sky forms an excellent source of illumination for the polarizing microscope.

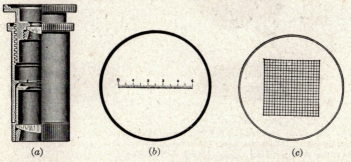

(a)   (b)   (c)

FIG. 21.—(*a*) Micrometer ocular; (*b*) scale in a micrometer ocular; (*c*) grating micrometer. (*Carl Zeiss, Inc.*)

In case such illumination is not available, artificial daylight lights can be successfully employed. These consist of various types of electric bulbs mounted in cases with a special blue-glass light filter in the path of the illumination. Three types are illustrated in Fig. 22. A low-voltage bulb with a condensing lens and diaphragm, as illustrated in Fig. 22c, provides suitable illumination for a wide variety of magnifications.

At high magnifications and for photomicrographic work a mechanical-feed arc lamp is sometimes used. The beam from

(a)

(b)   (c)

FIG. 22.—Various types of artificial illumination for the microscope: (a) small substage lamp; (b) strong lamp for general utility; (c) a low voltage light with a wide range of intensity. ((a), (b), *Spencer Lens Co.* (c) *Eric Sobotka.*)

the arc is very warm and should always be passed through a cooling cell of water in order to avoid injuring the cement in the prisms of the microscope (unless special prisms are employed).

Regardless of the source of illumination employed, it is important to regulate the light entering the microscope with respect to the optical system if the best results are to be achieved. In order properly to accomplish this result, suitable filters should be available for the source of illumination, the light used should be equipped with an iris diaphragm, and the condenser system should also contain a suitable diaphragm. The field of view in the microscope should be carefully bounded by each diaphragm and the proper filter system employed to reduce the illumination to suitable intensity. Proper resolution for each magnification may be secured in this way.

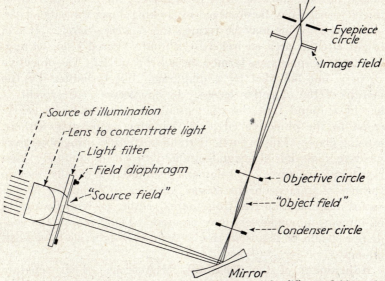

Fig. 23.—Diagram showing the relative dimensions of the different fields in the microscope and their relation to the illumination. (*After Belling.*)

The circular field of view seen by the observer when he looks through the microscope is bounded primarily by the diaphragm of the eyepiece. The diaphragm is fixed and also contains the crosshairs. The magnified image of this diaphragm bounds the *image field* seen through the eyepiece. The *object field* is a field of view equal in diameter to the diameter of the image field divided by the total magnification. It measures the area of the thin section or other object under observation at a particular instant. The *source field* is the field of view at the glass filter of the illuminator. The diameter of the source field is equal to the product of the diameter of the object field and the reciprocal of the reduction caused by the condenser. In the control of illumination, the area of the light leaving the illuminator should be cut by the condenser until it equals the source field. When the light entering the microscope is limited in this way, only the circle of the object that is seen is illuminated, and glare due to the interference of marginal light is eliminated.

When the condenser is in focus, the iris diaphragm determines the used aperture of the condenser. It is important that this aperture be filled with uniform illumination. If the objective is placed in focus and the eyepiece removed, the used aperture of the condenser may be observed by looking down the tube of the microscope. This may be termed the *condenser circle*. It is a bright circular area encircled by a dimly lighted band or ring. The latter is sometimes termed the *objective circle*. The objective circle is not bounded by a diaphragm but is limited by the margin of the objective lenses. In microscopic adjustment, it has been found that the condenser circle should be as nearly equal to the objective circle in diameter as possible without causing glare. This is particularly important in using objectives yielding high initial magnification with correspondingly high numerical apertures. Oil-immersion objectives usually require the use of immersion condensers in order to avoid the loss of useful magnification free from glare. Either corrected water or oil-immersion condensers may be used. The N. A. of the condenser should be less than the N. A. of the objective by a small amount.

Adjustment of the Polarizing Microscope.—Four separate steps can be outlined as necessary to arrange the polarizing microscope in order for the examination of rock sections:

1. Centering the stage with the field.
2. Crossing the nicols.
3. Testing the crosshairs.
4. Determining the vibration plane of the lower nicol.

1. *Centering the Stage with the Field.*—The stage is centered when the axis of rotation coincides with the tube axis of the microscope, the tube axis standing perpendicular to the center of the field of view. Screws on the side of either the objective collar or the stage (Fig. 24) are used to align the tube axis and the stage. A simple procedure is followed. While looking through the instrument at the field of view, pick out an easily recognizable point, and then rotate the stage. The point should

describe a concentric circle of rotation about the intersection of the crosshairs. If it does not, rotate the stage until the point is farthest from the intersection of the crosshairs, bring it in half-way by means of the centering screws, and then bring it to the center of the stage by actually moving the slide itself. Rotate the stage, and repeat the operation if the centering has not been completed the first time.

2. *Crossing the Nicols.*—The planes of vibration of the two prisms should be set at right angles to each other. The plane of vibration of the analyzer is usually fixed by the manufacturer either from left to right or up and down as one observes the microscopic field. The lower nicol is adjusted at right angles by

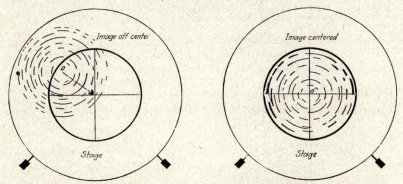

Fig. 24.—Diagram illustrating centering the field of view in the microscope.

rotating it in the substage collar until the field becomes dark, with both nicols in the path of light. The nicols should remain in the position giving maximum darkness. A small pin usually fits into a notch at this position.

3. *Testing the Crosshairs.*—The crosshairs in the ocular may be either the spiderweb type or lines engraved on a glass plate. In either case it is important that the hair lines be parallel to the planes of vibration of the two nicols. Ordinarily these are set by the optical firm supplying the microscope, and the ocular is so arranged that it will not fit the tube of the microscope in other than the correct position. The adjustment should be checked occasionally, however, and in case the alignment is inaccurate, the crosshairs should be reset by an experienced technician.

A slide containing small elongated rectangular crystals of natrolite (Fig. 25) is useful to test the setting of the crosshairs

with the planes of the nicols.[1]   The natrolite becomes dark
between crossed nicols when the edges of the crystals are parallel
to the vibration directions.   A slide containing a small natrolite

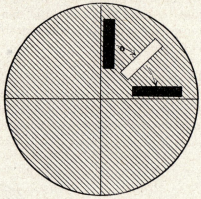

Fig. 25.—Testing the adjustments of the crosshairs with natrolite fragments.

crystal may be placed upon the stage between crossed nicols and
turned until it becomes dark.   If the crosshairs are in adjust-

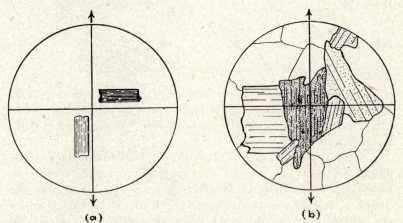

(a) (b)

Fig. 26.—Determining the vibration plane of the lower nicol: (a) elongated
tourmaline fragments; (b) biotite in thin section.

ment, the web lines should be parallel or at right angles to the
straight lines of the crystal.   This is true in each of the four

[1] If natrolite is not available, any crystalline material with straight-line
edges and parallel extinction may be substituted.

positions of extinction. In 45° intermediate positions the natrolite will show maximum illumination.

4. *Determining the Vibration Plane of the Lower Nicol.*—After the other adjustments have been made, the vibration direction of the lower nicol can be determined with either fibrous tourmaline fragments or a rock section containing biotite showing cleavage.

Tourmaline (Fig. 26a) has maximum absorption when it is oriented with the *c*-axis (usually the long direction of a crystal or fragment) in a direction at right angles to the plane of vibration of the polarizing prism. Biotite (Fig. 26b), on the other hand, is darkest when the cleavage is parallel to the vibration direction. Note the positions of greatest and least darkness, observing with the upper nicol thrown out from the tube. These indicate either the vibration direction or the normal to the vibration direction, depending upon whether the slide is biotite or tourmaline.

Suggested References

ALLEN, R. M.: "The Microscope," D. Van Nostrand Company, Inc., New York, 1940.

BECK, CONRAD: "The Microscope," R. & J. Beck, Ltd., London, 1938.

BELLING, JOHN: "The Use of the Microscope," McGraw-Hill Book Company, Inc., New York, 1930.

CHAMOT, E. M., and C. W. MASON: "Handbook of Chemical Microscopy," vol. 1, John Wiley & Sons, Inc., New York, 1930.

GAGE, S. H.: "The Microscope," Comstock Publishing Company, Ithaca, N.Y., 1925.

JOHANNSEN, A.: "Petrographic Methods," 2d ed., McGraw-Hill Book Company, Inc., New York, 1918.

MARSHALL, C. R., and H. D. GRIFFITH: "Introduction to the Theory and Use of the Microscope," Routledge, London, 1928.

SPITTA, E. J.: "Microscopy," E. P. Dutton & Company, Inc., New York, 1920.

Much useful information may be obtained from the catalogues of various optical firms:

Bausch and Lomb Optical Co., Rochester, N.Y.

E. Leitz, Inc., 730 Fifth Ave., New York.

Spencer Lens Co., Buffalo, N.Y.

Carl Zeiss, Inc., 485 Fifth Ave., New York.

CHAPTER III

A SUMMARY OF THE PROPERTIES OF LIGHT

Theories of Light.—Since light crosses interstellar space, penetrates transparent solids or liquids, and also travels through a vacuum, a medium has usually been postulated by which it could be conveyed. The medium is the ether, which has been assumed to permeate all matter and to pervade all space. Modern studies have shown, however, that in order to account for certain things the ether must be endowed with the most extraordinary physical properties, and according to some concepts it is unnecessary. The source of the light and its effect on the eye are apparent, but some explanation must be advanced to account for its transmission.

Several prominent theories have been advanced. According to one, a beam of light consists of a stream of minute particles, or "corpuscles," given off at high velocity by the sun or any luminous body. The corpuscles travel through space in straight lines and eventually reach the eye. This is generally referred to as the corpuscular theory, a theory that received much attention because it was supported by the famous physicist Sir Isaac Newton.

Another theory was first advanced by the Dutch scientist Christian Huygens in the latter part of the seventeenth century. According to Huygens, the ether is supposed to vibrate, and light is transmitted through it by the vibration of particle after particle in waves. The phenomena of light such as reflection, refraction, diffraction, and interference may be readily explained in accordance with this theory. The theory of Huygens, however, failed to explain the apparent rectilinear motion of light and was not accepted by Newton.

A modification of the wave theory was proposed by the Scottish physicist James Clerk Maxwell (1873), who considered light as made up of waves but said that the waves were electromagnetic. According to Maxwell, a wave consists of rapidly alternating electric and magnetic fields normal to each other and normal to the direction of propagation of light. Hertz (1888) succeeded in producing waves having properties similar to light waves by electricity. As a result of the work of Maxwell, Hertz, and other experimenters, the electromagnetic theory of wave motion was for a time generally accepted.

Toward the end of the last century evidence began to appear that did not accord with the electromagnetic theory. It was found that the space around certain metals would become electrically conductive when the metal was exposed to light. Then the electron was discovered in 1897, and it was assumed that the photoelectric effect was due to the emission of electrons as the metal became exposed to light. This was based on the fact that expulsion means energy, and it was presumed that the energy in the case of the photoelectric effect would come from light. However, the energy given by light is so small that it could not account for the emission of electrons. This led to the assumption that the light was concentrated in points and not uniformly distributed. At about this time Planck developed the assumption that radiating oscillators in a black body radiate energy discontinuously in units called *quanta*. Einstein in 1905 suggested that the absorption of light in the photoelectric process might also be in quantum units. Later experiment demonstrated that the quanta of Einstein were of the same size as those postulated by Planck.

As a result of these more recent developments the explanation of light seems to rest upon two contradictory theories, the wave theory being more appropriate for phenomena such as reflection, refraction, interference, diffraction, and polarization, whereas the quantum theory is more applicable to the recent discoveries in the field of X rays, radiation, and photoelectricity. Speaking of the two theories, Einstein has stated as follows:

We have good proof that both waves and particles exist. Our present effort is to understand how this is, to find a theory that will unify the nature of light. The composition of a two-point view has not yet been found. It is a quest of science in which our present methods are imperfect.

Nomenclature of the Wave Theory.—The nomenclature of the wave theory used in this text is summarized in the following paragraphs.

Vibration direction = electric vector = electrical displacement. The vibration direction lies in the wave front and is perpendicular to the ray in isotropic media. In anisotropic crystals it is not perpendicular except in limited directions.

Wave Front.—The surface determined at a given instant by all the parts of a system of waves traveling along the same direction and in the same phase. In space, in air, or in any other optically isotropic media when light moves along parallel lines the wave front is perpendicular to the direction of transmission. In anisotropic media the wave front is perpendicular only in certain directions.

Wave Length.—The distance between two successive crests or troughs, or any corresponding distance along the wave (denoted by the Greek letter lambda, λ). λ is usually measured in millionths of a millimeter ($m\mu$).

Beam.—A group of light waves following along the same path. A familiar example is the white beam of a motion-picture projector clearly visible in the dusty atmosphere of the theater. Beams can be made narrower and narrower.

Ray.—The limiting case of a beam is a single line and is called a ray. The ray is perpendicular to the electrical field and follows the direction of propagation of the energy. It indicates the direction of the transmission of the wave motion.

Refractive Index.—The refractive index is equal to the ratio of the wave-normal velocity in a vacuum to the wave-normal velocity (not the ray velocity) in the medium, whether isotropic or anisotropic.

Frequency.—The number of vibrations in a given unit of time. Ordinarily several trillion per second in the case of light waves.

Phase.—The relative position of corresponding points on different waves moving along the same line. Two points on waves are in the same phase when they are in the same relative position in regard to the crest or trough of the wave and are both moving either toward or away from the line of transmission. Two points are in opposite phase when they are in the same relative position but when they are moving in opposite directions with reference to the line of transmission. Other phase differences may occur.

The phasal difference represents the portion of a wave length by which one wave train fails to match the other.

Amplitude.—The maximum displacement of a wave from the line of transmission.

Period.—The time interval necessary for a wave to undergo a complete oscillation.

Crest.—The point of the wave with the maximum upward displacement.

Trough.—The point of the wave with the greatest downward displacement.

Monochromatic Light.—Light of a single wave length. In practical tests light is frequently used covering a small range of wave lengths but appearing as one color to the eye.

Light Vector.—The action of light may be described as depending upon the periodic alternation of a light vector that lies parallel to the plane of the wave front and in isotropic media is perpendicular to the direction of propagation. In anisotropic media the vector is still parallel to the plane of the wave front but, aside from certain limited positions, is not perpendicular to the direction of propagation.

In the case of monochromatic light the light vector follows the laws of simple harmonic motion, the vibration period T depending upon the color of the monochromatic light. The wave length λ—*i.e.*, the distance between two successive like points on a wave train—is equal to the velocity of propagation v multiplied by the vibration period:

$$\lambda = vT$$

In any transparent mass λ is fixed, and v varies with T. The intensity of light is the average of the intensities in the various light-vector quadrants and varies with the amplitude.

If the light wave is electromagnetic, there must be two vector movements in the system. In isotropic media these are transverse and perpendicular to each other, one limited by the magnetic field of force, the other by the electric field of force. These are connected with two other vectors distinguished in isotropic media by their magnitude and in anisotropic media also through variation in direction. These two vectors may be called the *electric vector* and the *magnetic vector*. The first of these measures the electrical displacement; the second, the magnetic displacement or induction. Maxwell has worked out equations applicable to the movement of these vectors (although the movement had been originally worked out before vector analysis was introduced). Experimental evidence has shown that the vibration direction of light corresponds to the electrical displacement (electric vector) in isotropic bodies. In anisotropic bodies it has been shown to correspond to either the electrical displacement (electric vector) or the electrical field. In the electromagnetic theory light is assumed to correspond to the electrical displacement (electric vector), an assumption made plausible by the conception of light as an electrolysis.

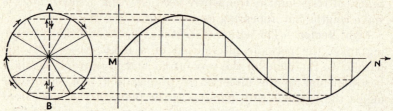

FIG. 27.—The harmonic curve produced by movement around a circle combined with motion along a straight line.

Speed of Light.—Light waves travel along the direction of transmission at a speed of approximately 186,284 miles per second. The same law of frequency used in the case of sound applies to light and is expressed by the equation

$$f = \frac{v}{\lambda}$$

or

$$v = f\lambda$$

In this equation the frequency f is obtained by dividing the velocity v by the wave length λ. In the case of violet light ($\lambda = 0.000037$ cm.) the velocity ($v = 186,284$ miles per second) divided by the wave length gives a frequency of 800,000,000,000,000 (eight hundred trillion) vibrations per second.

Wave Motion.—An idea of the behavior of light waves may be gained by a study of waves generated by simple harmonic motion and uniform rectilinear motion.

Simple harmonic motion is uniform motion in a circular path as it would appear projected on the diameter of a circle. If a particle as illustrated in Fig. 27 is assumed to move clockwise around the circumference of a circle, occupying various positions in turn, the projections on the vertical diameter AB will be at the intersections with the horizontal dotted lines shown in the figure.   If observed from the side along the plane of the circle, the particle will appear to oscillate back and forth with varying

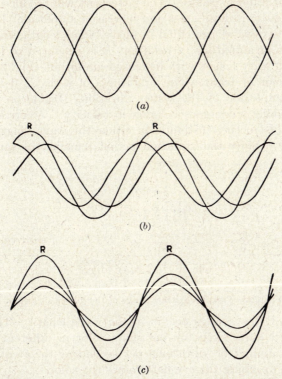

(a)

(b)

(c)

Fig. 28.—Combinations of waves produced by several phase differences.   (a) The phase difference is ½λ, and the two waves are equal and opposite in phase. (b) The resultant wave (R) produced by two equal waves of slight path differ- ence.   (c) The resultant wave (R) produced by two waves of equal length and identical phase, but differing in amplitude.

velocity.   If, in addition to the harmonic motion, the particle moves along a straight line MN at a uniform rate (rectilinear motion), it will no longer move in a circular path but will follow a curve of the type illustrated in the projection.   The projection on the vertical diameter of the circle, however, will still be the same.   The curve is a harmonic curve, which has the form of a sine curve.

Differences in phase produce a number of resultant forms when two or more waves follow the same line (Fig. 28). Two sets of waves may be equal and opposite, thus nullifying each other (Fig. 28a). Other sets may be equal in amplitude and wave length but differ in phase. The latter sort will produce a resultant R of intermediate crest and increased amplitude (Fig. 28b).

When two waves are equal in phase, wave length, and period but differ in amplitude, a resultant R is produced of the same phase and wave length with increased amplitude (Fig. 28c).

The Color of Light.—The brightness of a ray is determined by the amplitude of the wave vibration. Light, on entering various bodies, undergoes a change in velocity. A corresponding change must occur, therefore, in either the wave length or the frequency. Since the vibration period remains the same for a

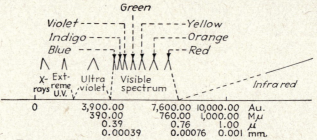

Fig. 29.—The approximate range of the visible spectrum.

given color, the change occurs in the wave length. The wave length will differ even for the same color in different bodies; thus the number of vibrations per second or the frequency of the waves reaching the eye determines the color.

Waves of visible light vary in length, the longest being red and the shortest violet. The portion of the spectrum directly visible to the eye varies between relatively narrow limits. The extreme wave length of red light is 0.0000759 cm., and the relative wave length of extreme visible violet is 0.0000393 cm. In terms of millimicrons (the units commonly employed in dealing with light), the figures are:

$$\text{Red light} = 759 \text{ m}\mu$$
$$\text{Violet light} = 393 \text{ m}\mu$$

White light, or ordinary light, is a combination of all the different wave lengths visible to the eye in one simultaneous effect. When only one wave length is observed, light is singly colored, or monochromatic. White light may be considered composed of seven different colors. These grade into each other, forming a continuous spectrum. The colors of the spectrum are frequently represented by arbitrarily chosen wave lengths representing mean values of the various colors, as follows:

$$
\begin{aligned}
\text{Red} &= 700 \text{ m}\mu \\
\text{Orange} &= 620 \text{ m}\mu \\
\text{Yellow} &= 560 \text{ m}\mu \\
\text{Green} &= 515 \text{ m}\mu \\
\text{Blue} &= 470 \text{ m}\mu \\
\text{Indigo} &= 440 \text{ m}\mu \\
\text{Violet} &= 410 \text{ m}\mu
\end{aligned}
$$

The electromagnetic spectrum extends far beyond the range of visible light. The mechanisms by which the different radiations are produced, however, must be much different because of the great difference in frequency.

Suggested References

COKER, E. G., and L. N. G. FILON: "A Treatise on Photo-elasticity," University Press, Cambridge, 1931.

CREW, H.: "The Wave Theory of Light," American Book Company, New York, 1900.

EDSER, E.: "Light for Students," Macmillan & Company, Ltd., London, 1930.

HARDY, A. C., and F. H. PERRIN: "The Principles of Optics," McGraw-Hill Book Company, Inc., New York, 1932.

HEYL, P. R.: The History and Present Status of the Physicist's Concept of Light, *Jour. Optical Soc. Am.*, vol. 18, pp. 183–192, 1929.

"Huygens' Treatise on Light," trans. by Silvanus P. Thompson, Macmillan & Company, Ltd., London, 1912.

NEWTON, SIR ISAAC, "Opticks," repr., McGraw-Hill Book Company, Inc., New York, 1931.

POCKELS, F.: "Lehrbuch der Kristalloptik," B. G. Teubner, Leipzig, 1906.

SAUNDERS, F. A.: "Survey of Physics," Henry Holt and Company, Inc., New York, 1930.

WEBSTER, D. L., E. R. DREW, and H. W. FARWELL: "General Physics for Colleges," D. Appleton-Century Company, Inc., New York, 1926.

WHITTAKER, E. T.: "History of the Theories of Aether and Electricity," Longmans, Green & Company, London, 1910.

CHAPTER IV

REFRACTION

Snell's Law. The Index of Refraction.—When light passes obliquely from one medium to another in which it travels with a different velocity, it undergoes an abrupt change in direction.

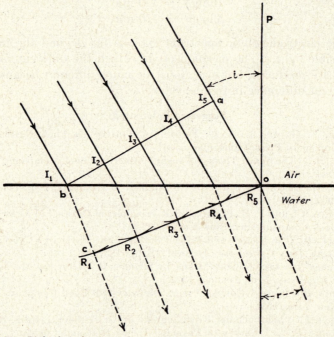

Fig. 30.—Light being refracted on passing from a rare into a denser medium.

This abrupt change in direction is known as *refraction*. The relationships of the incident and refracted light may be illustrated by the adaptation of the construction of Huygens shown in Fig. 30.

Let us suppose, for example, that a rare medium—air—is in contact with a denser medium—water. An incident beam *I* strikes the surface of the water obliquely, making an angle i with a perpendicular P. When the trace of the plane normal to the incident beam I strikes the surface at I_1, the point I_5 is still a considerable distance above the bounding plane. The positions I_2, I_3, and I_4, together with corresponding intermediate points, are also above the surface.

Let the beam advance until the ray at I_5 has reached R_5. During this advance the ray at I_1 has penetrated the denser medium and has continued with diminished velocity until it has arrived at the circumference of a circle with a radius I_1R_1, which represents the distance traveled in the denser medium. Similarly, I_2 has penetrated to the circumference R_2, I_3 to R_3, and I_4 to R_4. A tangent common to these circles represents the new wave front, and the new beam is perpendicular to the new wave front. The spherical waves sent out from b and other points on the bounding plane destroy each other except along bc and corresponding directions.

In the above construction the distances I_5R_5 and I_1R_1 may be considered proportional to the relative velocities of light in the two media.

It is apparent from the relationship of the lines of the diagram that

$$\sin i = \frac{ao}{bo}$$

or

$$bo = \frac{ao}{\sin i}$$

also

$$\sin r = \frac{bc}{bo}$$

or

$$bo = \frac{bc}{\sin r}$$

Since bo is common, the equations may be combined, and

$$\frac{ao}{\sin i} = \frac{bc}{\sin r}$$

or

$$\frac{ao}{bc} = \frac{\sin i}{\sin r}$$

The index of refraction is determined by the distance light will travel in a given time interval through a transparent substance as compared with air. In Fig. 30 light travels the distance *ao* in air, while it travels the distance *bc* in water. It follows, therefore, that the index of refraction

$$n = \frac{ao}{bc}$$

or

$$n = \frac{\sin i}{\sin r}$$

It appears from the foregoing equation that for any angle of incidence *the ratio of the sine of the angle of incidence to the sine of the angle of refraction is a constant.* It is also true that the respective velocities of light in the two media bear the same ratio. The relationship between the sines of the two angles and the velocities is known as *Snell's law.* It was discovered by Snell in 1621 but was not made known until after his death.

Let *n* be the index of refraction of a transparent material referred to air.[1] Then V = the velocity in air, and v = the velocity in the transparent material; also

$$n = \frac{V}{v}$$

If n_1 and n_2 are the indices of refraction of two different materials, then

$$\frac{n_1}{n_2} = \frac{v_2}{v_1}$$

Thus the indices of refraction of two transparent substances are inversely proportional to the velocities of light in the two media.

The angles *i* and *r* may be measured experimentally for many substances, thus determining *n*. The index of refraction depends both upon the substance and upon the kind of light. The indices of isotropic substances or general values are designated by the letter *n*. The extreme values for hexagonal or tetragonal

---

[1] The refractive index of dry air at 760 mm. pressure referred to a vacuum is only slightly different from unity (1.000274 at 15° C.); therefore indices of refraction of material substances referred to air are approximately equal to their indices referred to a vacuum (the latter are called the *absolute refractive indices*).

minerals are designated by n_ϵ and n_ω. Orthorhombic, mono-clinic, and triclinic crystals have their extreme values designated by n_γ (greatest), n_α (least), and n_β, the value in a direction at right angles to the two others.[1] The following table gives examples of values for the indices of refraction of several well-known minerals that occur throughout the normal range:

Minerals	(Na$_D$)	Indices of refraction
Fluorite.............	....	$n = 1.4338$
Quartz[1].............	....	$n_\epsilon = 1.5533$; $n_\omega = 1.5442$
Calcite[1].............	....	$n_\omega = 1.6585$; $n_\epsilon = 1.4863$
Apatite[1]............	....	$n_\omega = 1.6461$; $n_\epsilon = 1.6417$
Aragonite[1].........	....	$n_\alpha = 1.5301$; $n_\beta = 1.6816$; $n_\gamma = 1.6859$
Garnet (grossularite).	(Yellow)	$n = 1.7714$
Sphalerite..........	(Yellow)	$n = 2.3692$

[1] Quartz, calcite, and apatite are anisotropic with a range of values for refractive indices between n_ϵ and n_ω, the two extremes. Refractive indices of aragonite vary between n_γ and n_α.

Ordinarily, transparent minerals with a high index of refraction (1.9 or more) have the brilliant appearance called *adamantine luster*, while minerals with a lower index of refraction have a *vitreous luster*.

Dispersion.—The index of refraction for the violet end of the spectrum is greater than for the red end of the spectrum and on refraction red is deviated less than violet. The ordinary refraction of the two is indicated in Fig. 31.

The difference between the index of refraction for red and the index for violet is often briefly referred to as the dispersion. A

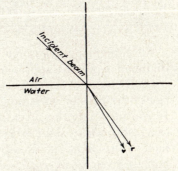

Fig. 31.—Variation in the angles of refraction for light of different colors.

cut prism placed in the path of a beam of white light produces a beautiful display of spectrum colors. The difference between

[1] In American technical journals dealing with optical descriptions of minerals many authors omit the letter n and use the Greek letters α, β, γ, ϵ, and ω alone in recording indices of refraction outside the isometric system.

the angle d_r and the angle d_v registers the dispersive power of the prism (see Fig. 39).

Minerals differ widely in their dispersive power. One having the least dispersion is fluorite; a mineral with one of the highest values is diamond. The "fire" or brilliant play of colors of the cut diamond is due to the high dispersion from the prismlike faceted gem stone. Fluorite, on the other hand, if cut and faceted, appears correspondingly dull. Because of its low dispersion optically clear fluorite is in demand for microscope lenses of high magnification to be corrected for chromatic variation. Comparative figures for the dispersion of fluorite and the diamond are given:

Fluorite		Diamond	
Illumination	Index of refraction, n	Illumination	Index of refraction, n
Red K(A'), $\lambda = 768.2$.....	1.43095	Red (B line)........	2.40735
Violet H(G'), $\lambda = 434.1$...	1.43963	Violet (H line)......	2.46476
Dispersion...............	0.00868	Dispersion..........	0.05741

All minerals have some dispersive power, but fluorite and diamond represent approximately the two extremes.

On account of the dispersion of minerals, accurate determinations of indices of refraction are made with monochromatic light. In routine study, however, the highest accuracy is seldom necessary, and white light is generally employed. As a matter of fact, white light as usually employed in determining mineral indices gives an average value for practical purposes somewhat comparable to yellow. The dispersive effect with white light is also extremely useful for several common tests.

Critical Angle.—In the formula $n = \sin i/\sin r$ the angle of incidence may vary between 0 and 90°. When $i = 0°$, the incident beam strikes the bounding surface at right angles. Sin i in this case $= 0$, and r must also equal 0. Thus an incident beam going from a rarer medium into a denser one is not refracted to either side but merely suffers a loss of velocity.

If $i = 90°$, $\sin i = 1$, and the equation becomes $n = 1/\sin r$. In this case, since n is a constant for a particular substance, the angle of refraction also becomes fixed. When $i = 90°$, the angle of refraction is known as the *critical angle*. The critical angle is important in the practical determination of indices of refraction.

Total Reflection.—Figure 32 shows light going from water into air (or from a denser medium into a rarer medium). A ray striking the surface of the water vertically from below continues out into air along the same path. If the path deviates from the vertical, the beam is refracted—*i.e.*, bent away—from the

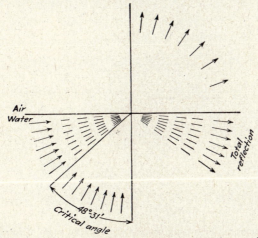

FIG. 32.—The relation between refraction of light passing from a dense medium into a rare medium and reflection beyond the critical angle.

perpendicular at the surface. When an angle of 48°31′ with the vertical is reached (the critical angle), the beam grazes the surface. For any angle greater than the critical angle, however, light is reflected downward. This phenomenon is known as *total reflection*.

If the angle with the perpendicular is increased until light travels along the surface between the air and water or strikes the surface at a grazing incidence, the beam is turned downward at the critical angle on the opposite side of the perpendicular to the bounding surface.

The same principle applies to all dense substances in contact with air. In practical determinations with a refractometer a glass hemisphere of high refractive index serves as the dense medium; light is directed against the hemisphere by a mirror, and the critical angle is determined with a measuring telescope.

Indices of Refraction of Anisotropic Minerals.—Optically, minerals belong to two classes: (1) isotropic and (2) anisotropic.

Minerals such as opal, glass, and other substances lacking regular internal structure and other minerals such as diamond, garnet, spinel, fluorite, etc., crystallizing in the isometric system are isotropic.   Minerals crystallizing in the other crystal systems are anisotropic.

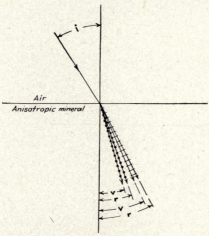

Fig. 33.—A diagram illustrating the variation of the angle of refraction for red and violet in an anisotropic mineral.

Light traveling through anisotropic minerals is doubly refracted as well as refracted.   Thus a beam of monochromatic light passing obliquely from air into an anisotropic medium not only is bent to one side but also is broken into two beams.   At the same time each of the two beams is polarized—*i.e.*, limited to a single plane of vibration, as will be explained in the chapter on polarized light.   In addition, each beam is differently refracted for different colors of light.

Double refraction of two rays occurring in anisotropic minerals is illustrated in Fig. 33.   The incident beam is broken into two sets of rays, white light producing two dispersed spectra with opposite directions of vibration.   Monochromatic red light will yield two angles of refraction, and monochromatic violet light will also yield two different angles. In each case of monochromatic light and also in the case of white light the directions of vibration are opposed but may not be exactly at right angles in the mineral. On emerging into air on the opposite side of a mineral plate, however, the vibration is at right angles.

Measurement of Indices of Refraction by Refractometers.— Several types of refractometer have been devised for determining the indices of refraction of liquids or of solids. A glass hemisphere is utilized in different ways.

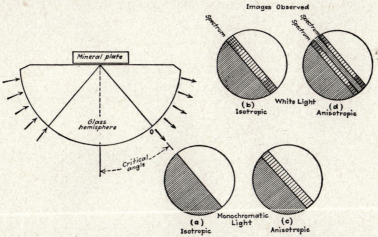

Fig. 34.—Illumination from below in determining the critical angle by total reflection.

Figure 34 illustrates the determination of the index of refraction by the method of total reflection. Light is directed against the surface of a glass hemisphere from below at an angle of refraction greater than the critical angle. The light rays are reflected downward from the upper surface of the hemisphere and emerge on the opposite side. The material to be determined, either a mineral plate with a polished lower surface or a liquid, is placed on the hemisphere. If the material is a mineral plate, a thin film of oil of high refractive index is placed between the hemisphere and the mineral. The light from the mineral is reflected through the hemisphere in part and produces an image in the observing telescope focused at *o*, as illustrated in the figure. The upper half of the field will be dark, while the lower half is illuminated (or *vice versa* if the image is inverted). If monochromatic

light is used and the mineral is isotropic, the boundary between
the light and dark areas will be sharp, marking the critical angle
accurately (Fig. 34a).   On the other hand, in the case of white
light the colors of the spectrum lie *between* the light and dark
areas (Fig. 34b).   If the mineral is anisotropic and the observation occurs with monochromatic light, two boundary lines will
be seen spaced a short distance apart (Fig. 34c).   If an anisotropic mineral is observed with white light, the two boundaries
will be marked by spectra (Fig. 34d).   When the anisotropism
is not strong, the two spectra frequently overlap.   The light

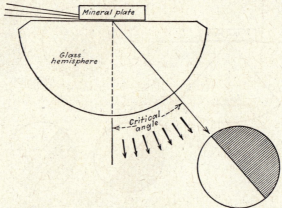

Fig. 35.—Determination of the index of refraction with the incident beam grazing
the surface of the hemisphere.

waves of each image from the anisotropic mineral are polarized
at right angles to the other.

Instruments operating on this principle are arranged with
graduated scales and verniers for measuring the critical angle.
When the critical angle is known, the index of refraction is computed from Snell's law, sin i being equal to 1.   The wave length
is controlled at the source by using monochromatic light.

The method of grazing incidence can be applied with the same
glass hemisphere used for the method of total reflection.   Light
enters the mineral from the side and is refracted downward
through the hemisphere as shown in Fig. 35.   In this case the
upper half of the field of view is dark, and the lower half is
illuminated.   Otherwise the same principles prevail that apply
to the other method.   The contrast between the two fields is

more pronounced when the mineral is illuminated by grazing incidence.

AB = double prism in cooling chamber	Q = clamp
C = illuminating prism	R = illuminating mirror sending light to AB
E = cooling tube; water circulates in direction of arrow	
F = telescope tube	S = sector with engraved scale
J = holding arm for movable sector	T = turning gear
L = reading lens	Th = thermometer
M = adjustment for compensator	V = adjustment screw
Oc = eyepiece	Z = compensator scale

FIG. 36.—The Abbe refractometer. (*Carl Zeiss, Inc.*)

Whether employing total reflection or grazing incidence, the index of refraction of the glass hemisphere must be accurately known. This is usually ascertained by determining the index of refraction by total reflection in reference to air with the mineral plate removed, before determining the index of refraction of the mineral. The index of the liquid used for mounting the mineral plate can also be determined by total reflection if a few drops are smeared on the upper surface of the hemisphere. The refractive

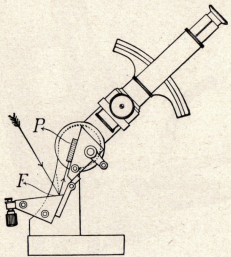

Fig. 37.—Sectional view of the refractometer. (*Carl Zeiss, Inc.*)

index equation for the determination of the mineral may be stated:

$$n \text{ (mineral)} = n \text{ (glass)} \times \sin r$$

where n (mineral) = index of refraction of the mineral
n (glass) = index of refraction of the hemisphere
r = critical angle for the mineral as read with the refractometer

The glass used in constructing hemispheres has a high index of refraction, usually about 1.80. The mineral should have a polished surface to be placed toward the hemisphere, and a liquid with a high index of refraction is pulled by capillary attraction between the mineral and the hemisphere. Methylene iodide ($n = 1.74$) is usually preferred for this purpose.

Figure 37 is an illustration of one of the standard Abbe (Zeiss) refractometers commonly used in determining the indices of refraction of liquids (also of minerals and crystals). The refractive index is read off a graduated sector, and the instrument may be quickly set for reading directly. A scale is arranged in the image of the eyepiece that gives the values of indices of refraction for all angles within the range of the instrument, which extends from $n_D = 1.3$ to $n_D = 1.7$.

A refractometer of this type consists essentially of a double prism *AB* that receives the liquid to be tested, a telescope for viewing the line of the critical angle, and a scale sector *S* for reading the angle. The double prism is made up of an illuminating prism *F* (Fig. 37) with a ground-glass surface and a refracting prism *P* that operates on the principle of grazing incidence. The refracting prism takes the place of the glass hemisphere in other refractometers. A compensating device is inserted in the telescope to approximate monochromatic light (D line of sodium). The refractometer is usually adjusted for use by testing the instrument with a glass plate of known index.

Index of Refraction by the Prism Method.—In Fig. 38 a beam of monochromatic light strikes the prism *ABC*. At *AB* the beam

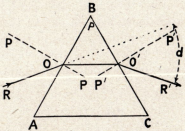

R is bent toward the perpendicular *OP*. At *BC* the beam *R'* emerges and is bent away from the perpendicular *O'P'*. *R'* continues in a straight line from this point, making an angle of deviation *d* with the original direction.

When the angle *d* is the minimum that can be observed when the prism is turned with respect to the beam (the angle of minimum deviation) and the angle *p* of the prism is also known, the index of refraction of the glass of the prism may be computed. The formula by which the index is computed is

Fig. 38.—A beam of light bent to one side by passing through a glass prism.

$$n = \frac{\sin \frac{1}{2}(d + p)}{\sin \frac{1}{2}p}$$

p = angle of the prism
d = angle of minimum deviation of the beam
n = index of refraction of the prism

When the beam of light striking the prism is not monochromatic but consists of the various wave lengths that combine to produce ordinary light, a spectrum is produced as shown in Fig. 39.

The prism formula is particularly useful for determining the indices of refraction of glass prisms, transparent crystals, and hollow prisms filled with oil. A prism is adjusted vertically on a one-circle goniometer and the prism angle measured by obtain-

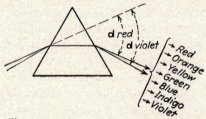

FIG. 39.—Diagram illustrating the variation in the angle of minimum deviation of a prism with the wave length of light.

ing reflections first from one side and then from the other and reading the angle on the graduated circle. With the same set-up a beam of light is passed through the prism and the angle of minimum deviation is measured. The index of refraction is then computed directly from the formula given above, utilizing the angle of the prism and the angle of minimum deviation.

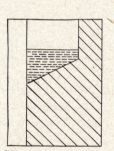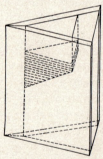

FIG. 40.—Glass prism to be mounted in a goniometer for determining the indices of refraction of liquids. (Hollow space for liquid in upper portion of wedge between the plates.)

Hollow prisms made of specially ground thin glass or plates with parallel surfaces or selected glass slides cut and cemented together are useful for determining the indices of refraction of liquids. Figure 40 illustrates a hollow prism made of glass plates of equal thickness and having parallel surfaces. Bakelite may be used to cement the glass plates to a solid prism. In ordinary determinative work, one prism cut at 45° is usually sufficient. Two prisms, one cut at 30° for high determinations and one at 60° for lower values, may be employed. Occasionally microscopic slides will possess the parallelism of surfaces required for the prism walls. Such slides should be tested by observing the reflection of a hanging window cord drawn taut by a suspended weight or some other suitable straight line from both surfaces of the slide. The slide should be held close to the eye to observe this and turned until the reflections from both the upper and the lower surface can be seen at the same time. If the straight lines are parallel and uniformly distant in each image, the slide is satisfactory. It should also be tested in two perpendicular positions. When a good slide is found, it should be cut into two pieces, one for each side of the prism. The two parts are beveled and cemented[1] to a solid glass wedge. If the walls of the hollow prism are made of glass with parallel sides, correction for the glass is not necessary. It is best, however, to assume an exaggerated bevel and mount the two sides in opposed positions. Any existing lack of parallelism will be largely corrected in this way. Index of refraction determinations using this method are useful for determining the indices of refraction of liquids beyond the range of commercial refractometers. The method is also suitable when a refractometer is not available and as a check when it is available.[2]

If the same prism is always used, a chart may be prepared giving the index of refraction corresponding to each angle of minimum deviation for a given wave length. If the chart covers the range of indices of refraction from 1.400 to 1.850, it will include all ordinarily determined values. Indices of refraction may be determined with the light of a sodium flame obtained by holding an asbestos sheet previously saturated in salt in a fan-flame burner. Light from a mercury-vapor arc or white light transmitted through a standard color filter is also occasionally employed. A helium-gas tube gives an ideal sodium line.[3]

---

[1] Bakelite resinoid, baked at 70° C. for 10 hr. and followed by baking at 125° for 10 hr., will make a solid prism. Bakelite varnish containing china oil is affected by index liquids and should be avoided.

[2] Hollow prisms of several types have been developed by Dr. E. S. Larsen of Harvard University, Dr. C. S. Ross of the U. S. Geological Survey, and Dr. H. E. Merwin of the Geophysical Laboratory.

[3] Employed in the laboratories of the U. S. Geological Survey.

One of the best methods of securing a sodium flame has been reported by F. Lowell Dunn, M.D., of Omaha, Nebr. He uses a *coarse* alundum filtration crucible filled with salt and suspends it over a Méker burner. A molten hemisphere of NaCl forms

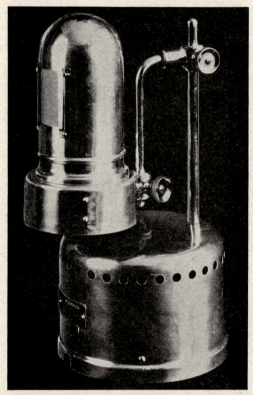

Fig. 41.—Sodium laboratory arc furnishing a strong source of sodium light. (*Nela Specialty Division, Lamp Department of General Electric Company.*)

in the bottom of the crucible, which gives a sodium flame of extremely high intensity. The burner should be placed under a hood and at a safe distance from the microscope.

Special light bulbs giving a strong sodium light have been developed by the General Electric Company (Fig. 41). These require about 20 min. to acquire the proper color value but after developing the correct intensity furnish an excellent source of illumination.

The refractive indices of the various types of glass used in optical equipment are determined to the eighth decimal place with sodium light, mercury-vapor light, and several other light sources. Such precision is not employed in examining minerals with the microscope, nor is it possible without a special goniometer.

The Determination of the Index of Refraction with the Microscope.—The index of refraction of a mineral is seldom determined completely in examination of thin sections. The slices are mounted in Canada balsam, and the usual test consists in ascertaining whether a given mineral has an index of refraction greater or less than balsam. The indices of adjacent minerals are also compared with each other.

The indices of refraction of adjacent transparent substances can be compared in several ways. The two most useful methods are the method of central illumination and that of oblique illumination.

A method of direct determination with the microscope exists, but unfortunately it is not sufficiently accurate with thin sections.

The Method of Central Illumination.—The test is best made with a magnification of 80 or greater, with the iris diaphragm partly closed. It is quite sensitive to small differences in refractive indices at such magnifications. If monochromatic light is employed, it is possible to distinguish between the indices of refraction of two minerals even when they differ by as little as one in the third place of decimals.

The test may be applied to thin sections of transparent minerals in comparing their relative indices of refraction with adjacent minerals or balsam. The phenomenon used in making the test depends upon the total reflection of light incident at more than the critical angle when passing from a mineral of greater to a mineral of lesser index in a thin section. The test is employed for comparing the indices of refraction of the various minerals of thin sections with balsam, for comparing the minerals with each other when observed in contact, and for comparing fragments of minerals with various immersion media in which they may be mounted. Light enters the section from below and is transmitted through both media. At the bounding edge both reflection and refraction take place, and a portion of the entering beam is bent either to one side or to the other, depending upon the relative indices of refraction of the adjacent media. If the two indices happen to be the same, no refraction takes place. In case the index of refraction of one is greater than that of the other, light will strike an inclined boundary between the two in some place at an angle greater than the critical angle. A portion of the beam will be deflected toward the mineral with the greater index. If the boundary between the media is not inclined, grazing incidence may occur, bringing about the same effect. The deflection results in a light blur, visible through the microscope just inside the boundary of the mineral grain of greater index. The blur is more apparent if the iris diaphragm is partly closed and if the tube of the microscope is slightly raised. It forms an irregular white line; and as the tube is raised still farther, an illusion is produced, the line appearing to move

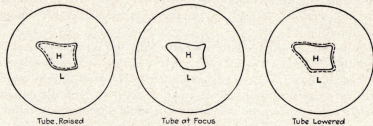

Tube Raised   Tube at Focus   Tube Lowered

Fig. 42.—Central illumination, $n >$ balsam.

toward the center of the mineral. If the tube is lowered, the effect is reversed.

Since the index of balsam is known (approximately 1.537), minerals may be quickly divided into two groups, one with indices greater than balsam, the other with indices less than balsam. It is convenient to remember in making the test that when the tube is *raised*, the line moves toward the medium having the *higher* refractive index. Conversely, when the tube is *depressed*, the line moves toward the medium having the *lower* index (Fig. 42).

Explanation of the White-line Effect.—Hotchkiss has given an explanation of the refraction and reflection involved in the method of central illumination. The construction shown in Fig. 43 is modified from his explanatory diagram.

Two minerals *A* and *B* are assumed to be in contact in a thin section with a vertical bounding plane *YZ*. *A* is the mineral with a lesser index ($n = 1.50$), and B has a greater index, 1.70. A cone of light rays enters the two minerals from the balsam below, divided evenly on both sides of the bounding plane. The cone of light may be represented by the rays 1, 2, 3, and 4 with angles of inclination as indicated in the diagram.

The critical angle in mineral B with respect to the bounding plane is 61°55′. The half cone of light within B (3 and 4) strikes the surface YZ at an angle greater than the critical angle and is totally reflected. The half cone of light within A (1 and 2) is split at the bounding surface, part being refracted into B and part being reflected back into A. The comparative intensity of

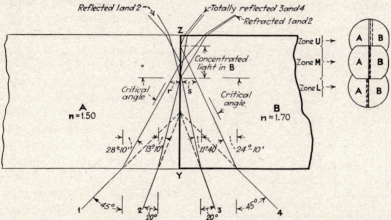

FIG. 43.—A theoretical explanation of the movement of the light line in central illumination. (*Modified from Hotchkiss.*)

the reflected and refracted rays depends upon the character of the bounding plane. If the contact surface is highly polished, more light is reflected and less refracted. If the surface is rough, as is usually the case, more light is refracted into B.

A zone of light is concentrated within the mineral of higher index at level M. Within this vertical distance nearly all the light of the cone is concentrated in B. If the plane of focus of the microscope is brought within this zone, a band of light is visible within the mineral with a higher index. If the plane of focus is elevated by raising the microscope tube to level U, the band becomes broader and furnishes the illusion of moving toward the center of B and away from the bounding plane. If the plane of focus is lowered to level L, on the other hand, a greater con-

centration of light is present in *r* than in *s*, and the light band will
appear to be within the mineral with a lower refractive index.
The circles adjacent to the braces and indicating the vertical
extension of the zones are intended to illustrate the positions of
the white line corresponding to different elevations of the plane
of focus.

Oblique Illumination.—The method of inclined or oblique
illumination is more convenient for making the same relative
comparisons of refractive indices outlined above, but at a
magnification of about 50 or lower. A larger area of the thin
section is included within this field, and the method allows the
observer quickly to compare a large number of mineral grains;
also, it provides an easier inter-
polation of values between two
mounting media.

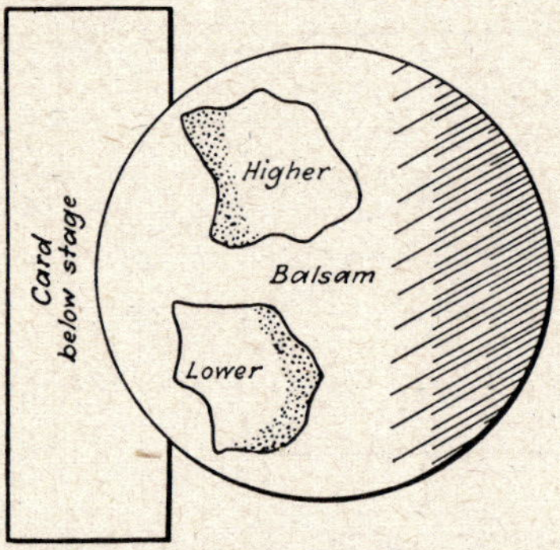

Fig. 44.—Illustrating oblique
illumination with the condenser
removed. Higher = mineral grain,
n > balsam. Lower = mineral
grain, *n* < balsam.

The effect is best observed
without a condenser lens.
Oblique illumination may be secured by inserting a card below
the stage cutting off half of the
light. This darkens one-half of
the field, at the same time allow-
ing the opposite half to be illuminated largely by oblique rays
(Fig. 44). A similar effect may
be secured by inserting a narrow
card in the accessory slot above
an objective of moderate power
(with the condenser in the system). This effect may be either
the same or reversed, depending upon the focal length of the
condenser.

Individual crystals of minerals will be unevenly illuminated
by the method of oblique illumination. One side of the mineral
will be dark, and the opposite side will be light. When the
card is inserted below the objective, the shadow will appear
either on the side of a mineral toward the dark half of the field
or on the side away from it. When the shadow appears on the
side away from the dark half, the index of refraction of the mineral
in question is greater than that of the adjacent medium; if on the
side next to the dark half, the index of the mineral is less. In

case the index of refraction of the mineral is about equal to the index of the mounting material and white light is employed, one side will be blue while the other is red. When the card is inserted in the accessory opening, the shadow in the mineral is on the side of the field next to the dark area if the index of the mineral is greater than balsam.[1]

The index of a known mineral should be tried first in making this test in order to be sure of the set-up in the microscope.

Double Diaphragm Method.—Saylor's investigation of the sensitivity of various methods of matching indices of refraction with the microscope has led to the proposal of the double-diaphragm method, a modification of the method of inclined illumination. This is more particularly applicable to the immersion method, as described in Chap. VIII. The set-up, with a mineral fragment mounted in an immersion liquid of higher index, is shown in Fig. 45. The solid lines show the light rays entering the mineral, and the dash lines indicate emerging rays.

Relief.—Certain minerals stand out strongly in the field of the microscope, others are moderately visible, and frequently the mineral is hardly visible at all. This appearance or visibility of outline and surface is described as *relief*.

The relief of a mineral mounted in balsam depends upon the difference between the index of refraction of the mineral and balsam. Minerals with low indices of refraction (cryolite, $n = 1.364$) and high indices of refraction (spinel, $n = 1.75$) have strong relief. On the other hand, such a mineral as apophyllite has approximately the same index of refraction as balsam and consequently is hardly visible in thin section.

Anisotropic minerals with a wide divergence between the two extremes of refractive indices exhibit a variation in relief as the stage of the microscope is rotated. Calcite furnishes one of the best illustrations of this feature. The ray vibrating parallel to the short diagonal of the cleavage rhombohedron has nearly the same index as balsam. When this direction is parallel to the lower nicol, a calcite cleavage fragment on the stage of the microscope shows low relief. When the cleavage fragment is turned at right angles until the long diagonal is parallel to the

---

[1] Another simple way to secure the effect of oblique illumination suggested by Dr. J. D. H. Donnay consists in shading the field by partially inserting the frame containing the analyzing nicol.

lower nicol, light travels through the mineral with the velocity of the higher index, and the same grain stands out with high relief. A number of common minerals vary in relief with direction.

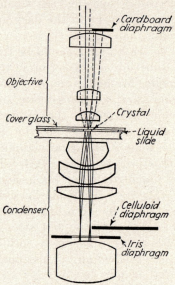

Fig. 45.—Double-diaphragm method of oblique illumination.   (*C. P. Saylor.*)

The relief of a mineral may be estimated as *low, moderate, high,* or *extremely high.*   In the tables to follow, such a descriptive term is given for each mineral.

Suggested References

BECKE, F.: *Sitzb. Akad. Wiss. Wien*, C11, Abt. 1, pp. 358–378, 1893.

GRAHAM, G. W.: *Mineralog. Mag.*, vol. 15, pp. 341–347, 1910.

HOTCHKISS, W. O.: *Am. Geol.*, vol. 36, pp. 305–308, 1905.

SAYLOR, C. P.: Accuracy of Microscopical Methods for Determining Refractive Index by Immersion, *Nat. Bur. Standards, Research Paper* 829, vol. 15, pp. 277–294, 1935.

WRIGHT, F. E.: The Methods of Petrographic-microscopic Research, *Carnegie Inst. Pub.* 158, 1911.

CHAPTER V

PLANE POLARIZED LIGHT IN MINERALS

Polarized Light.—In the foregoing it has been assumed for descriptive purposes that light may be considered as wave motion. This condition holds for ordinary white light or for monochromatic light of any sort. It is also assumed that the vibrations take place in all directions around the line of transmission. Under certain conditions, however, the tendency to vibrate in all directions around the line of transmission is modified, and the waves become restricted for the most part to a single direction of vibration. When its vibration direction is thus restricted, light is said to be *polarized*.

Polarization of light may be brought about in several ways: (1) by reflection from a polished surface; (2) by repeated refraction at an angle through several plates of thin glass; (3) by absorption by certain crystals such as tourmaline or herapathite; (4) by specially constructed prisms of optically clear calcite.

Polarization by Reflection.—
Light reflected obliquely from a polished surface, such as a table top, is partially polarized. If the reflection is examined through a rotating nicol prism, the field of view in the prism

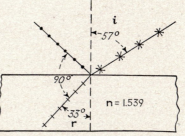

FIG. 46.—Polarization by reflection.

will darken whenever the vibration plane of the nicol is at right angles to the plane of reflection of the polished surface.

Light reflected and refracted obliquely from mirrors is partially polarized. According to Brewster, the polarization in the case of a glass plate is a maximum when the directions of reflection and refraction are 90° apart (Fig. 46). When these two directions are at such an angle, the angle r becomes the complement of the angle i, and the formula $n = \sin i/\sin r$ may be written $\sin i/\cos i = \tan i = n$. Thus, at the angle of maximum polarization the tangent of the angle of incidence equals the index of refraction of the reflecting substance. Consequently, when the index of refraction of the substance is known, the angle of maximum polarization may be obtained from a table of tangents. The angular relationships for a plate $n = 1.539$ are shown in Fig. 46. Figure 47 is a sectional view of an old-fashioned polariscope employing reflection from glass plates to obtain polarized light. The instrument was used before the advent of the modern polarizing microscope to produce polarized light for the study of mineral plates.

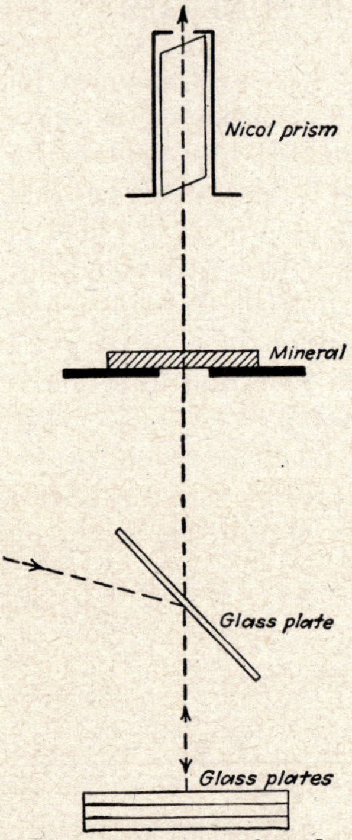

Polarization by Absorption.—Tourmaline has the property of producing polarization by absorption. Light that strikes the crystal vibrating in a variety of planes is strongly absorbed except along one plane. As a result, the rays of light that emerge are

Fig. 47.—Polarization by reflection in a polariscope.

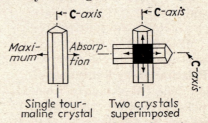

Fig. 48.—Darkness due to absorption produced by two superimposed tourmaline crystals.

limited to this plane of vibration and are thus plane polarized. The crystallographic axis c, frequently the long direction of the crystal, lies parallel to the plane of vibration.

Observation through either a Nicol prism or another plate of tourmaline cut in a similar fashion effectively reveals the polarization. When the plane of the nicol is at right angles to the optic axis of the tourmaline plate, the crystal appears dark. Also, when the directions of the two superimposed tourmaline plates are at right angles to each other, the overlapping portion is dark (Fig. 48).

Thin crystals of a strongly absorptive compound, iodo-cinchonidine-sulfate, were described in 1852 by William Bird Herapath, M.D. (Fig. 49). Because of their strong absorption in one direction corresponding to the behavior of tourmaline, the crystals were referred to as "artificial tourmalines." The material was subsequently called *herapathite* in honor of the discoverer. More recently, methods have been developed for producing thin transparent sheets containing small crystals of herapathite in parallel orientation embedded in a plastic binder. The effect corresponds in a general way to that of a single crystal of large area, and a polarizing sheet results. Two overlapping sheets of this material are illustrated in Fig. 50. It is possible to prepare such polarizing sheets covering several square feet in area.

Fig. 49.—Crystals of herapathite showing an area of extinction where individuals with directions of greatest absorption at right angles are superimposed. (*After Herapath*, 1853.)

Fig. 50.—Two discs of herapathite ("Polaroid") mounted on glass plates photographed with the planes of polarization at right angles. (*Courtesy of the Polarizing Instrument Company.*)

Up to the present time, because of low absorption in the violet, polarizing sheets have not been substituted for optically clear calcite in the polarizing microscope. Such sheets are excellent, however, for optical demonstration and for many supplementary uses, particularly where a wide field of polarized light is required.

Double Refraction (Birefringence).—Although polarization results when light is transmitted through most transparent minerals, the phenomenon is not ordinarily accompanied by absorption. With a few exceptions, light in passing through transparent minerals is doubly refracted into two beams vibrating along two planes that are approximately at right angles to each other.[1] This applies except in the case of amorphous minerals and minerals crystallizing in the isometric system.

The most obvious illustration of double refraction and accompanying polarization by a mineral occurs in transparent calcite,

FIG. 51.—Double refraction illustrated by a cleavage rhomb of transparent calcite or Iceland spar.

or Iceland spar. Objects viewed through a cleavage plate of Iceland spar appear double; if a cleavage face of calcite is placed over a dot within a circle marked on a piece of paper, the dot will appear to the eye as two dots and the circle as two circles (Fig. 51). The light giving rise to one image will be composed of waves vibrating parallel to the long diagonal; that giving rise to the other will be composed of waves vibrating parallel to the short one. The two light rays have been differently refracted, and the indices of refraction are different.

The cleavage form of calcite, the rhombohedron, is illustrated in Fig. 51 with the principal axis in the vertical position. If the opposite vertices of the calcite cleavage having threefold symmetry are ground to triangular surfaces and polished, light may be passed directly through parallel to the principal crystallographic axis [perpendicular to (0001)]. The light rays vibrate

[1] F. E. Wright has demonstrated that the precise determination of the angle between the two rays is a matter of careful physical measurement. In discussing double refraction, the amount of variation from 90° will not be taken into account, and the two rays will be considered in simple terms as about at right angles.

uniformly about this axis, and there is an absence of double refraction. This is the optic axis in either hexagonal or tetragonal crystals and agrees in direction with the principal crystallographic axis (the *c*-axis).

If light passing along the optic axis is examined by means of a Nicol prism, it is found that there is no double refraction and the mineral appears isotropic. In any other direction, however, double refraction results. In the latter cases light is polarized into two rays vibrating at right angles to each other, one vibrating at right angles to the optic axis, the other in a plane through the optic axis. The former is known as the *ordinary ray;* the latter

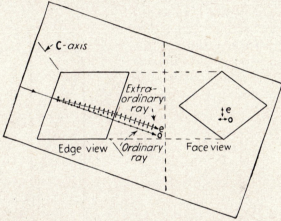

Fig. 52.—Edge and face views of a cleavage rhombohedron of calcite illustrating double refraction.

is called the *extraordinary ray.* The ordinary ray vibrates in a direction at right angles to the optic axis and parallel to the long diagonals of the rhombic faces of the cleavage rhombohedron as shown in Fig. 52. The extraordinary ray vibrates in a plane passing through the optic axis and also passing through the short diagonal. In some minerals the extraordinary ray is the fast ray; in others it is the slow ray.

Nicol Prism.—The Nicol prisms in the polarizing microscope utilize the principle of double refraction to produce polarized light. Optically clear calcite is used, and a prism is made of two parts cemented together with Canada balsam. The two halves form a prism of the type illustrated in Fig. 54. Light entering the base of the prism is broken into extraordinary and ordinary rays. The extraordinary ray has an index of refraction $n = 1.516$ at the angle of incidence for the prism; the ordinary ray has an index of refraction $n = 1.658$. The index of the extraordinary ray is close to the index of refraction of balsam, $n = 1.537$. The index of the ordinary ray, however, is considerably greater. Both rays strike the cementing plane of balsam obliquely. The obliquity of the ordinary ray exceeds the critical angle between the ordinary ray and balsam. As a result, it is not refracted through the balsam but is reflected to the side of the prism. Since the extraordinary ray does not exceed the

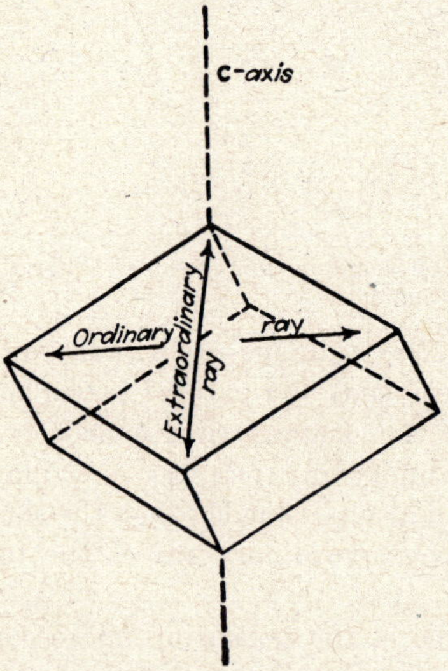

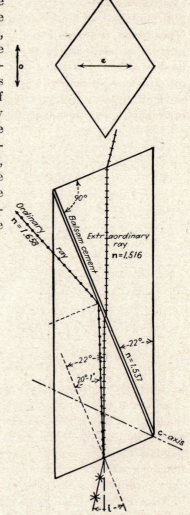

FIG. 53.—The extraordinary and ordinary rays in a calcite cleavage.

FIG. 54.—The polarization and deviation of light in a Nicol prism.

critical angle between the extraordinary ray and balsam, it passes on through the prism with little deviation.

The extraordinary ray is polarized with one plane of vibration; consequently, the light emerging from the prism and made up entirely of the extraordinary ray is plane polarized.

Interference between Crossed Nicols.—When two Nicol prisms are superimposed with their planes of vibration at right angles to each other, the nicols are said to be crossed. The polarizing microscope is normally adjusted with the Nicol prisms in this position, the plane of each nicol remaining fixed but the upper nicol movable either in or out of the tube of the microscope. Crossed nicols produce darkness when the stage is unoccupied or when it holds optically isotropic materials such as glass or opal or crystals of the isometric system of crystallization. Minerals crystallizing in crystal systems other than the isometric are anisotropic and in most positions produce interference colors between crossed nicols.

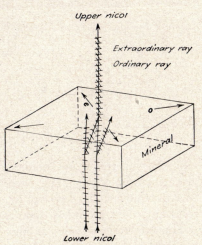

FIG. 55.—The vibration directions of the extraordinary and ordinary rays in an anisotropic mineral illuminated with polarized light.

In Fig. 55 polarized light is shown passing through a mineral plate after leaving the lower nicol. Light strikes the lower surface of the mineral plate vibrating in one plane. On entering the plate, it is broken into two sets of rays. Both sets of rays are polarized, and light travels with different velocities along them. As a result, when the two sets of rays emerge on the upper side of the plate, one set has traveled farther than the other, and extraordinary rays will travel along the same direction as ordinary rays, both vibrating almost at right angles and having traveled different distances. Both travel along a straight line to the upper nicol and continue to vibrate at right angles.

The location of the planes of vibration is determined by the position of the mineral. If the position of the mineral plate on the stage is changed as shown in Fig. 56, the vibration planes of the emerging rays are also shifted.

After leaving the mineral plate, the two rays of the mineral, *e* and *o* in Fig. 57, continue to the analyzer. Here one of the rays *e* is broken into two components *eo'* and *ee'*, the ordinary ray *eo'* being refracted to the side of the nicol, the extraordinary ray *ee'* continuing through the upper half of the nicol. The separation of the rays follows the same principle already explained in the discussion of the Nicol prism. The other ray *o*, from the mineral, is also broken into two rays in the analyzer. One component *oo'* is reflected to the side, and the other *oe'* continues along the plane of the analyzer. As a result of this selection, two rays in the upper half of the analyzer *ee'* and *oe'* emerge as extraordinary rays

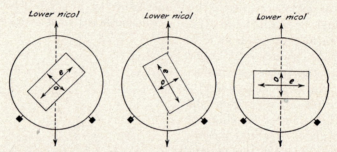

Fig. 56.—A mineral plate on the microscope stage showing several positions of the doubly refracting rays.

vibrating in the same plane. Although these two rays are vibrating in the same plane, each has traveled a different distance. In consequence, the two are in a position to interfere, and the resultant effect produced upon the eye is an interference color.

The interference color produced depends upon the nature of the light and the amount of retardation of one set of waves with respect to that of the other. The retardation can be determined and is expressed by the Greek letter Δ. The value of Δ is expressed in millimicrons (millionths of a millimeter = mμ), the same units used to measure the wave length of light.

The retardation may be changed through a wide range by (1) varying the thickness *t* of the mineral, (2) changing the orientation in such a way as to change the indices of refraction n_1 and n_2 of the two rays emerging from the mineral. This relationship may be expressed by the equation

$$\Delta = t(n_2 - n_1)$$

In the equation, t represents the thickness of the mineral expressed in millimeters, n_2 is the greater index of refraction, and n_1 is the lesser index of refraction for a particular orientation.

Phase Difference.—The two rays emerging from the mineral differ in phase, or in other words they have a phase difference P. This difference is equal to the retardation divided by the wave length:

$$P = \frac{\Delta}{\lambda}$$

Since it has just been shown that

$$\Delta = t(n_2 - n_1)$$

it follows that

$$P = \frac{t(n_2 - n_1)}{\lambda}$$

When the retardation is some whole multiple of a wave length ($n\lambda$), the waves emerging from the upper nicol become equal and opposite in phase. The resultant is then equal to zero, and the field produced is dark (Fig. 58).

When the retardation is $\left[\dfrac{(2n + 1)}{2}\right]\lambda$, the components of the waves in the plane of the upper nicol are equal and on the same side of the line of transmission. The resultant wave is equal to the sum of the two components, and maximum intensity results (Fig. 59).

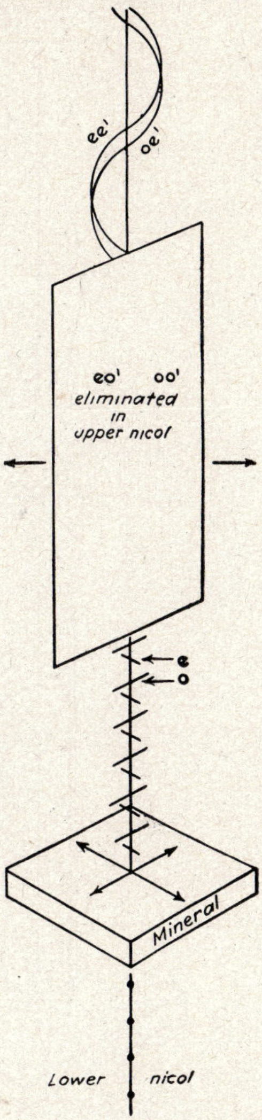

Interference Colors.—If the mineral plate lies with the planes of vibration parallel and perpendicular to the planes of the two nicols, no light passes through the analyzing nicol, and the mineral is in a position of extinction. On the other hand, if

FIG. 57.—Sorting of rays by the upper nicol when the nicols are crossed.

the plate is rotated to either side, the field of the upper nicol
is no longer dark but becomes illuminated with interference

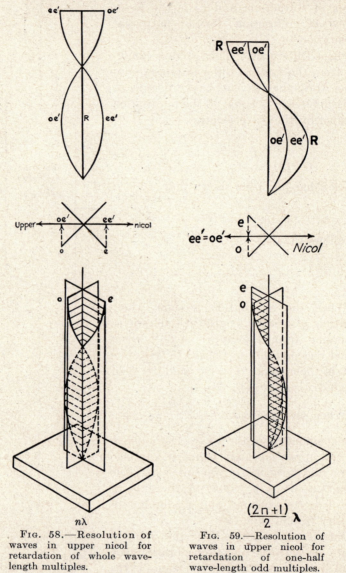

FIG. 58.—Resolution of
waves in upper nicol for
retardation of whole wave-
length multiples.

FIG. 59.—Resolution of
waves in upper nicol for
retardation of one-half
wave-length odd multiples.

colors. The interference colors vary with the thickness of the
mineral section, the nature of the mineral, the way in which

the mineral section is cut, and the light employed.   The explanation of the relationship of these various factors involves many of the principles of optical mineralogy.   It is desirable for the sake of simplicity to consider the variables one at a time.

If the thickness of a mineral plate between crossed nicols is changed, the orientation remaining the same, a change in

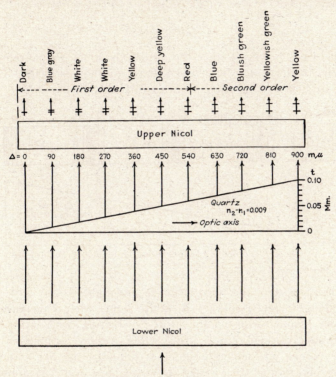

Fig. 60.—Interference colors due to a portion of a quartz wedge between crossed nicols with white light.

interference color ensues.   One of the best ways to illustrate this phenomenon is by means of the quartz wedge that accompanies the polarizing microscope.

Figure 60 is a diagram illustrating a portion of a quartz wedge cut along the *c*-axis and varying in thickness from 0.0 to 0.10 mm. The wedge is placed between crossed nicols in a position at 45° to the planes of the nicols.   In this position it becomes brilliantly illuminated with interference colors.   The colors, however, gradually merge into each other and change as one observes different thicknesses along the wedge. Any one thickness, however, forms a uniform band of one color across the wedge. The quartz wedge should be placed on the stage of the microscope and moved back and forth in order to observe the full range of color due to varying thickness.

Each portion of the wedge is subject to the equation

$$\Delta = t(n_2 - n_1)$$

In this case, however, since the optic axis of the wedge remains parallel to the stage, $(n_2 - n_1)$ is fixed and equals 0.009. Consequently, the retardation Δ varies with the thickness t.

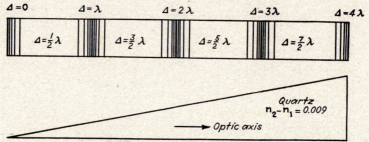

FIG. 61.—Alternate dark and light bands produced by monochromatic light with a quartz wedge between crossed nicols.

When t is zero, the retardation in any light is also zero, and the field of view is dark. In white light, when t increases, a definite sequence of colors ensues. Starting with gray and continuing through bluish gray, white, yellow, orange, in the order named, the colors become striking to the eye. In the thicker portion of the wedge, however, less contrast appears; and in wedges several times as thick the colors at the thick end become faint iridescent tints.

If the source of illumination is changed and monochromatic light is used in the system, a different effect is produced, as illustrated in Fig. 61. In this case, when the thickness reaches such a point that the retardation becomes equal to one wave length, the two monochromatic waves are equal and opposite in phase and nullify each other, causing darkness. As a result, dark bands will occur at all points where the retardation is a whole multiple of λ. Conversely, at odd multiples of $\frac{1}{2}\lambda$, maximum intensity will occur. Here the two waves are equal and in the same phase.

The interference colors due to white light are a subtraction of all the various wave lengths of the spectrum from white. The method by which the various interference colors are related to their monochromatic components is illustrated by Fig. 62.

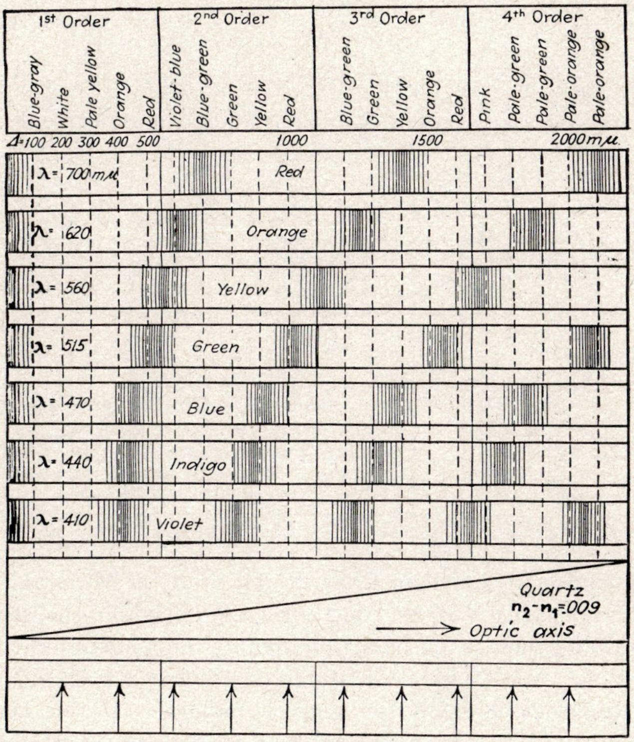

Fig. 62.—The relationship between interference colors due to monochromatic light and colors due to white light.

The various monochromatic beams, on passing through a wedge, produce dark bands at different thicknesses. Likewise, maximum intensity occurs at corresponding intermediate intervals. The difference between the wave lengths at the opposite end of the spectrum is such, however, that the first dark band for violet occurs almost in the first position of maximum intensity for red. For violet the band is approximately 410 mμ. Since the wave length for red is about 700 mμ, the maximum intensity for red occurs at 350 mμ ($\frac{1}{2}\lambda$). When the thickness and double refraction are such that the retardation equals 410 mμ, no violet is present in the resultant interference color. The percentage of maximum intensity for red at the same time is about 83. Since the maximum intensity for red occurs at $\frac{1}{2}\lambda$ or 350 mμ, the percentage intensity at 410 mμ would be

$$\frac{2(\lambda r - \lambda v)}{\lambda r} \times 100 = \frac{2(700 - 410)}{700} \times 100 = 83 \text{ per cent}$$

If the wave lengths are known, it is possible to compute the percentage of any given monochromatic light present in an interference color of a given retardation.

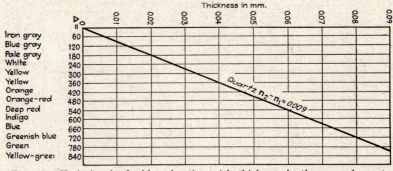

Fig. 63.—Variation in double refraction with thickness in the case of quartz.

Application of the Color Chart to the Study of Minerals.—The *interference color chart* is constantly employed in the study of minerals by means of polarized light. The maximum double refraction, or the greatest difference between n_2 and n_1, is approximately constant for a given mineral. If this constant is substituted in the equation $\Delta = t(n_2 - n_1)$, a straight-line curve is the result. In the case of quartz, for example,

$$(n_2 - n_1) = 0.009$$

If various thicknesses are assumed, as shown in Fig. 63, the corresponding retardation Δ may be determined directly. If the normal sequence of colors for a given retardation is known, it is possible either to predict the color of quartz of a given thickness or to tell the thickness of quartz having a given interference color, provided the quartz is in such a position that $n_2 - n_1$ is a maximum.

This relationship not only is true for quartz but may be applied, with the exception of a few special instances, to all anisotropic minerals studied with the petrographic microscope. The color chart (facing page 163) gives the lines of maximum double refraction for the common minerals.

In the color chart interference colors with Δ less than 550 mμ are said to belong to the *first order*. Violet ($\Delta = 550$) belongs at the boundary of the first order. This is known as *sensitive violet*, since a small change either way produces a decided color difference. From violet $\Delta = 550$ to violet $\Delta = 1128$ the colors belong to the second order. From violet $\Delta = 1128$ to violet $\Delta = 1652$ they belong to the third order. Above the fourth order colors are not easily separated. The colors at the end of the first order and the beginning of the second are the most striking and brilliant. At the end of the fourth order they merge into each other, forming tints of green and pink tending toward grayish white. These colors are quite distinct from the blue gray, white, and yellowish white of the lower first order. Uncertainty concerning the order of a given color may be overcome by using a mica plate. The mica plate is cut with such thickness that it increases or decreases retardation of a section by about $\frac{1}{4}\lambda$ (sodium light). Such an increase or decrease in the lower first or second orders produces a set of colors entirely different from that in the case of a similar change in higher orders. For example, in the case of first-order yellow $\Delta = 400$ mμ, an increase in Δ of 175 mμ will result in violet $\Delta = 575$ mμ, and a decrease of the same amount will produce white $\Delta = 225$ mμ. The same increase or decrease in retardation above the fourth order would produce little change perceptible to the eye.

Determination of Retardation with a Berek Compensator.— M. Berek (1913) described a rotary calcite compensator of simple mechanical construction. A calcite plate 0.1 mm. thick, cut normal to the optic axis, rests on a rotating axis in a metal holder similar to the frame usually used for the gypsum and mica plates. The frame is held fast in the accessory slot of the microscope by a spring. The rotation of the compensator plate is registered on a graduated drum attached to the axis of rotation. The drum is graduated with a vernier reading to tenths and may be calibrated in degrees.

The plate in the compensator is held in a small ring that may be easily removed, and a plate of different thickness may be substituted. The range of the plate ordinarily employed covers retardations from zero to the fourth order.

The axis of rotation of the compensator is arranged diagonally to the polarization planes of the two nicols. If the planes of the nicols are north-south and east-west, the tube slot holding the compensator will be northwest-southeast. The compensator is

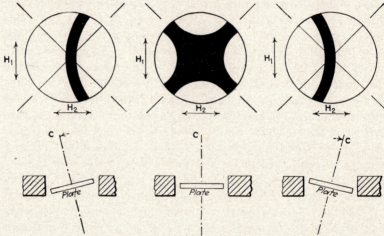

Fig. 64.—The vibration directions and movement of the color rings in the field of the ocular when using the Berek compensator.

marked with two arrows: H_1, parallel to the axis of rotation or along the accessory slot, is the slow-ray vibration direction; H_2, at right angles to the axis of rotation, indicates the trace of the projection of the plane containing the inclined *c*-axis of calcite and marks the fast-ray vibration direction.

The compensator is first set with the plate horizontal within the frame and inserted. Between crossed nicols a large dark cross will appear in the field. When this cross coincides with the crosshairs of the microscope, the compensator is in the zero position (see Fig. 64). If the compensator drum is then turned either to the left or to the right, the various orders of interference colors appear in the field in a sequence corresponding to the order of the quartz wedge.

The compensator may be used to determine the retardation of a mineral grain between crossed nicols as follows: The grain in question is moved to the center of the field and placed in the 45° position with the slow-ray vibration direction of the mineral parallel to H_2 of the compensator. The compensator is then inserted and rotated first to the right and then to the left, stopping in each case when the interference color of the mineral has been completely reduced to extinction. The measured difference between the opposite readings is divided by two and the value inserted in a simple formula supplied by the makers of the

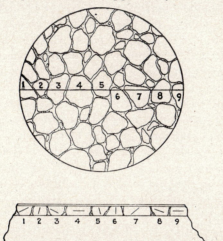

FIG. 65.—Determination of thickness of section in quartzite.

instrument. Solution of the formula gives the correct retardation for the mineral grain.

A view of the Berek compensator is shown in Fig. 17. Figure 64 indicates the views obtained in the microscope field with the compensator plate horizontal and rotated either to the right or to the left. The vertical sections in the lower part of Fig. 64 indicate the inclination of the c-axis, and the upper diagrams represent corresponding microscope fields. With monochromatic light, light and dark bands are produced on either side of a central cross. With white light, the bands on either side of the dark cross indicating the zero position are colored.

When the compensator is inserted above a doubly refracting crystal in a thin section, the dark cross disappears. As the plate is rotated, however, the interference colors are changed until complete compensation occurs as mentioned above.

Determination of Thickness of Section.—Let us suppose that Fig. 65 represents a thin section containing numerous small quartz grains in random orientations. For purposes of illustration we shall refer to grains 1 to 9 inclusive along the horizontal crosshair in the field of the microscope. These grains are oriented with their optic axes lying in the positions shown in the sectional view. Most are inclined; occasionally a few are vertical and a few are horizontal. The horizontal axes are in the correct position to provide a maximum value of $(n_2 - n_1)$. Since all are of uniform thickness, the grains with horizontal axes will show the highest order of interference color, which is likewise the interference color with maximum retardation. In any section of uniform thickness that has a large number of grains, as in the case illustrated, the grains giving the highest order interference color as observed by means of the color chart will be grains in a position to exhibit the maximum $(n_2 - n_1)$. In the case at hand, grain 4 is in the correct position. If grain 4 should show an interference color of straw yellow, the thickness of the section as determined by the color chart would be 0.03 mm. Other interference colors will be shown in the thin section, but only those with an approximately horizontal position will be as high in the first order as straw yellow.

In any thin section, if sufficient grains of a known mineral are present in random orientation and the highest order of interference color can be determined, it is possible to ascertain the thickness of section by reference to the color chart. It is also possible to reverse the process if the thickness is known and determine the double refraction of an unknown mineral. Likewise, in a slide containing two or more minerals, one of which is known, it is possible to determine the thickness of the section from the known mineral and determine the double refraction of the unknown minerals from the determined thickness and the observed interference colors. These considerations are extremely useful in studying minerals with the polarizing microscope.

Direction of the Vibration of Slow or Fast Rays.—It is frequently important to ascertain the planes of vibration of the two rays vibrating at right angles in an anisotropic mineral grain. The two rays have different indices of refraction, the one with the greater index being the slow ray and the one with the lesser index, the fast ray. The determination of the fast- and slow-ray directions is accomplished between crossed nicols, the location of the two rays being established by observing the position of extinction. When the mineral becomes dark, the vibration directions of the two rays are parallel to the planes of vibration of the Nicol prisms. Since the planes of vibration of the nicols are parallel to the crosshairs in the ocular, the vibration planes in the mineral will also be parallel to the crosshairs when in the extinction position.

A mica plate or a gypsum plate is used to tell which of the two rays is fast and which is slow. When the positions of the vibration directions of the rays are ascertained, the mineral is turned from extinction to the position of maximum interference. Next, either the gypsum or the mica plate is inserted in the tube of the microscope with the slow-ray vibration direction parallel to one of the vibration directions of the mineral. If the order of color increases, the parallel direction is the slow-ray vibration direction of the mineral. If it decreases, the direction represents the fast ray. One direction being known, the other is the opposite. The mica plate is usually used for minerals with weak double refraction, and the gypsum plate is employed in the case of stronger double refraction. When the mineral has very strong double refraction, a quartz wedge may be used. Since the quartz wedge varies in retardation from zero to the fourth order, a variety of colors will be produced, the color at a particular part of the wedge depending upon the thickness. When the slow ray coincides with the slow-ray direction in the mineral, a corresponding reinforcement in retardation will occur. Thus the color of the mineral will suddenly change to a color of higher order, dependent upon the portion of the wedge superimposed. When the slow-ray direction in the wedge is opposed to the slow-ray direction in the mineral, subtraction occurs.

Extinction.—When a mineral plate or grain or a portion of a doubly refracting crystal is dark between crossed nicols, it is said to lie in the position of extinction. Frequently minerals have prominent cleavage lines or crystal boundaries that enable one definitely to locate the angle at which extinction occurs with respect to the crystallographic axes. In the absence of a reference line, the extinction angle cannot be determined.

Parallel Extinction.—Frequently minerals have a single plane of cleavage. The traces of the cleavage planes appear in thin sections as irregularly spaced parallel lines. If the mineral becomes dark between crossed nicols, with the cleavage parallel to the vibration directions of the two nicols, the extinction is said to be parallel.

A number of minerals crystallize in such a way that sections are elongated, square, or rectangular. Square or rectangular cleavage patterns may also be observed. If these minerals become dark between crossed nicols, with the cleavage directions parallel to the vibration planes of the nicols, they are said to have parallel extinction.

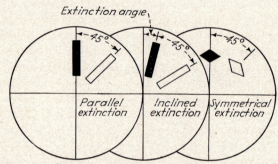

Fig. 66.—Relative positions of greatest and least illumination in parallel, inclined, and symmetrical extinction as observed between crossed nicols.

Inclined or Oblique Extinction.—Many minerals extinguish between crossed nicols when cleavages or crystal boundaries lie at oblique angles to the planes of vibration of the two nicols. These are said to have inclined extinction.

In this case it is necessary to know the position of either the fast-ray vibration direction or the slow-ray vibration direction in the mineral grain. The extinction angle is usually determined in terms of the slower of the two rays, or the one having the greater index of refraction. The nature of the two rays is determined with one of the accessory plates of the microscope.

Several different angles of extinction are usually observed for the same mineral in a given section, as illustrated in Fig. 67. The maximum reading on the slow-ray vibration direction with the plane of vibration of the analyzer is a convenient value to determine. In the case of observation with the microscope, the stage is rotated until the mineral lies in a position of extinction. The upper nicol is then pushed to one side, and the angle between the vertical crosshair (parallel to one of the nicols) and the cleavage line or crystal boundary is determined by readings on the graduated stage of the microscope. The nicols are then crossed again and the crystal turned to the extinction position, the angle being measured. Next, the direction of vibration of the slow

Fig. 67.—Diagram illustrating various positions of an elongated mineral with a maximum extinction angle of 51° on the slow ray as it might appear in thin section.

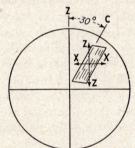 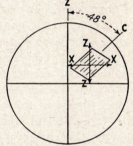

Fig. 68.—Hornblende of Fig. 264*b* in the position of maximum extinction. between crossed nicols.

C ∧ Z = −30°.

Fig. 69.—Hedenbergite of Fig. 245*b* between crossed nicols in the position of maximum extinction.

C ∧ Z = −48°.

ray is verified by using an accessory plate. A series of readings should be repeated with different crystals until it seems certain that the largest angle for a particular mineral has been found. When the angle is determined, it is necessary to refer to a description of the optical directions in the crystal in order to ascertain the proper reference plane for the extinction angle.

The mineral descriptions in Part II of this text include the angles of extinction. The angle between Z and the *c*-axis of a crystal is frequently recorded. Since Z is a slow-ray direction and prominent cleavages or crystal boundaries are often referred to the *c*-axis, it is usually possible to interpret the extinction from the orientation diagram.   Figures 68 and 69 furnish illustrations of such interpretations.

Symmetrical Extinction.—A number of minerals form cleavage patterns or crystals with rhombic cross sections.   In many instances these become dark between crossed nicols when the planes of vibration of the nicols are parallel to the diagonals of the rhombic patterns.   Extinction of this type is described as symmetrical.   Several minerals forming crystals with square outlines may also yield symmetrical extinction.

Elongation.—Occasionally crystal grains develop with an elongated habit and straight edges.   These may have a lathlike shape under the microscope, may resemble small needles, may occur in long crystals, or may show several other shapes of similar development.

When such crystals are anisotropic, it is possible to determine the fast- and slow-ray vibration directions with one of the marked accessory plates.   In case the vibration direction of the slow ray of the crystal is parallel to the long direction, the mineral is said to have *positive elongation*.   When the vibration direction of the slow ray lies across the crystal in the short direction, the mineral has *negative elongation*.   These two terms may be stated briefly as *length-slow* and *length-fast*, length-slow indicating that the vibration direction of the slow ray is parallel to the length of the crystal, and length-fast indicating the parallelism of the vibration direction of the fast ray.

Anomalous Interference.—Occasionally minerals normally assumed to be isotropic become anisotropic and give interference effects between crossed nicols.   The abnormal production of interference colors often of a low order is called *anomalous*. Figure 121 represents a thin section of garnet that exhibits symmetrically arranged bands of interference colors photographed between crossed nicols.   X-ray studies show that the same garnet is still isometric in crystallization, so the colors are truly anomalous.

Interference colors and structural patterns may be produced by strain in the crystals.   According to Crookes, the great Cullinan diamond, measuring almost 4 in. across, exhibited pronounced anisotropism due to strain.

Idocrase in thin section often shows an unusual sequence of interference colors, Berlin blue predominating.   Although this mineral is tetragonal and normally doubly refracting, the interference colors do not follow the color chart and are anomalous. Clinozoisite, zoisite, brucite, and some varieties of chlorite furnish other examples of anisotropic minerals that yield anomalous interference colors.

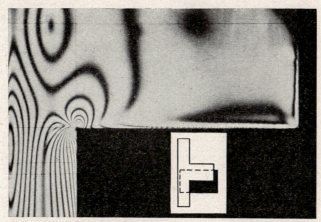

Fig. 70.—Photograph between crossed nicols of equal interference areas in strained bakelite cut in the form of a structural T and placed under pressure. (*Courtesy of Photo Elastic Laboratory, Department of Civil Engineering, Columbia University; photograph by Raymond D. Mindlin.*)

Equal interference areas are frequently produced in isotropic bakelite through strain. In Fig. 70 a portion of a small bakelite frame cut in the form of a T is shown between crossed nicols. The T would have a shape illustrated by the insert, the portion photographed being outlined by the dotted lines. The photograph was obtained by utilizing monochromatic green (5461 A.U.) in the mercury spectrum.

Suggested References

BEREK, M.: Zur Messung der Doppelbrechung hauptsächlich mit Hilfe des Polarisationsmikroskops, *Centl. Mineral.*, pp. 427–435, 1913.

BOUASSE, H.: Optique cristalline double refraction polarisation rectiligne et elliptique, 1925.

DRUDE, PAUL: "Theory of Optics," trans. by Mann and Millikan, Longmans, Green & Company, New York, 1925.

GROTH, P.: "The Optical Properties of Crystals," trans. by B. H. Jackson, John Wiley & Sons, Inc., New York, 1910.

HARTSHORNE, N. H., and A. STUART: "Crystals and the Polarizing Microscope," Edward Arnold & Co., London, 1934.

JOHANNSEN, A.: "Manual of Petrographic Methods," McGraw-Hill Book Company, Inc., (1918). A summary of the various types of polarizing prisms will be found on pp. 158–175.

MACCULLAGH, JAMES: Crystalline Reflexion and Refraction, *Trans. Roy. Irish Acad.*, vol. 18, pp. 31–74, 1837.

MIERS, H. A.: "Mineralogy," 2d ed., rev. by H. L. Bowman, Macmillan & Company, Ltd., London, 1929.

SCHUSTER, A., and J. W. NICHOLSON: "Theory of Optics," Edward Arnold & Co., London, 1924.

TUTTON, A. E. H.: "Crystallography and Practical Crystal Measurement," 2d ed., vol. 2, Macmillan & Company, Ltd., London, 1922.

WEINSCHENK, E.: "Petrographic Methods," trans. by R. W. Clark, McGraw-Hill Book Company, Inc., New York, 1912.

WINCHELL, A. N.: "Elements of Optical Mineralogy," 5th ed., Part I: Principles and Methods, John Wiley & Sons, Inc., New York, 1937.

WRIGHT, F. E.: The Transmission of Light tnrough Transparent Inactive Crystal Plates, etc., *Am. Jour. Sci.*, 4th ser., vol. 31, pp. 157–211, 1911.

The student is referred to comments by Tunell and Morey regarding certain fundamental optical properties, *Am. Mineralog.*, vol. 17, pp. 365–380, 1932.

CHAPTER VI

CONVERGENT POLARIZED LIGHT

General Statement.—The lens combination used in the microscope for obtaining interference figures is usually described as conoscopic. The usual arrangement produces interference figures visible in the field of the ocular. Such figures are particularly useful for determining the optical directions in crystals. Their interpretation involves the principles outlined in the preceding chapter on polarized light, combined with an understanding of the crystallization of minerals.

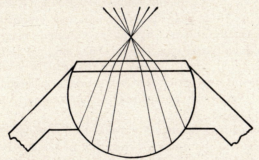

FIG. 71.—Convergent light produced by the front lens of the condenser.

In obtaining interference figures of small crystals, it is necessary to exercise particular care to have all of the elements in the optical train exactly aligned and properly centered. It is best to use a moderately high magnification, preferably a 4-mm. objective, although an 8-mm. objective is sometimes satisfactory and may be more easily manipulated. The auxiliary condenser should be inserted across the axis of the microscope below the stage. The front lens of the condenser throws a concentrated convergent beam against the mineral plate (Fig. 71). Some microscopes are also provided with a diaphragm between the polarizer and the lower component of the condenser. The diaphragm limits the field of view and helps to improve the outer portion of the interference figure. A Bertrand lens is inserted in

the tube of the microscope above the analyzer.   This lens brings
the image of the interference figure into focus on the focal spot
of the ocular.   The figure then becomes visible to the observer
through the ocular.   Good figures of small size can be obtained
by removing the ocular and not using the Bertrand lens.   A black
disc with a small hole in the center, or any one of several appli-
ances designed for this purpose, may be used to replace the ocular
when an interference figure is obtained without the Bertrand
lens.

Two types of interference figures are given by anisotropic
minerals: uniaxial and biaxial.   Minerals crystallizing in the
hexagonal and tetragonal systems are uniaxial; those crystallizing
in the orthorhombic, monoclinic, and triclinic systems are biaxial.
The difference between uniaxial and biaxial minerals is fundamental and is due to the arrangement of the atoms set up in
crystallization.   Occasionally biaxial crystals have such a small
axial angle as to appear uniaxial, and conversely on certain
occasions normally uniaxial crystals may become biaxial because
of strain.   Such variations should be examined with caution
when encountered, yet they need not disturb the student's con-
fidence in the interpretation that tetragonal or hexagonal crystals
are uniaxial, whereas monoclinic, orthorhombic, and triclinic
crystals are biaxial.

Formation of Interference Figures.—Convergent light passing
through a crystal plate causes variation in retardation between
crossed nicols.   This variation in retardation has many points in
common with the variation in retardation obtained with the
quartz wedge, as described in the discussion of parallel polarized
light.   The use of the quartz plate instead of a wedge and of
convergent polarized light instead of parallel polarized light
produces interference colors dependent upon the convergence of
the beam.   The varying angle of illumination of the oblique
rays results in varying values of n_2 and n_1 for a doubly refracting
mineral.   Varying values of n_2 and n_1, in turn, cause varying
retardation.

When a quartz plate is being examined the most striking
interference effect occurs with the optic axis of the plate at right
angles to the microscope stage.   The same fundamental considerations that have been demonstrated to hold true in the case of the
wedge also apply to the plate.   Here, however, the thickness

remains constant, and the double refraction $(n_2 - n_1)$ varies with the retardation, depending upon the direction. The angle of incidence on the quartz plate due to the convergent beam employed varies from 0 at the center of the field to a maximum on either edge. As a result, the difference $(n_2 - n_1)$ also changes from 0, at the center where the incident beam is parallel to the

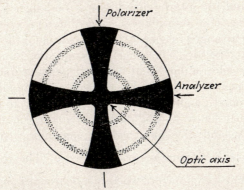

FIG. 72.—A uniaxial interference figure looking down on an optic axis.

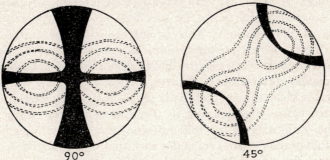

90°   45°

FIG. 73.—A biaxial interference figure in 90° and 45° positions.

optic axis, to considerably greater values at the edge of the field. Darkness or total extinction occurs at the center of the field and where the vibration directions of the inserted plate are parallel to the vibration directions of the nicols, resulting in a black cross for quartz. The explanation lies in the fact that convergent light strikes the surface of a mineral plate not only along a straight line, as in a section of a quartz wedge illuminated by parallel polarized light, but also radially around the center. Consequently, vibration directions will be arranged tangentially

and radially throughout 360° of rotation.   As a result, vibration
directions of the extraordinary and ordinary rays from the plate
will be parallel to the vibration planes of the nicols in certain
directions.   The two directions are directions of extinction and
in general uniaxial minerals form dark cross arms at 90° (Fig. 72).
In biaxial minerals the positions of extinction show greater
variation, and the interference figure is no longer a simple cross
but changes as shown in Fig. 73.   The different orders of color in
the field are arranged in concentric circles around the center of the
cross.   Other factors remaining the same, the number of color
bands observed in a particular field is dependent upon the thickness of the plate and the double refraction of the mineral.

Monochromatic light produces alternate dark and light bands
in interference figures.   The dark bands correspond to retardations of $n\lambda$, and the intermediate maximum colored bands
correspond to a retardation of $\dfrac{(2n+1)\lambda}{2}$.   The relationship is
similar to that which results when monochromatic light is passed
through a quartz wedge.   The colors in interference figures
produced by white light are actually a combination of the different monochromatic wave lengths due to the varying oblique
angle of illumination.   This is analogous to the interference
color chart where white light results as a summation of the various
monochromatic wave lengths due to variation in thickness.

Uniaxial Interference Figures.—Hexagonal and tetragonal
minerals yield the characteristic axial cross of a uniaxial interference figure when viewed in the direction of the optic axis.
If the optic axis of the mineral (the same in direction as the
crystallographic c-axis) coincides with that of the microscope, the
uniaxial figure will be centered with the two arms crossing at
the intersection of the crosshairs in the microscope.

However, if the optic axis is inclined to the axis of the microscope, the point of intersection of the cross arms will fall away
from the intersection of the crosshairs.   It frequently falls outside the field of the microscope.   If the center of the axial cross
does not coincide with the center of the field, the point of intersection of the arms will move around the crosshair intersection
when the stage is rotated, describing a circle and returning to
its original position after rotating 360°.   The intersection of the
cross arms marks the point of emergence of the optic axis, and

its deviation from the center of the field is a measure of the angle between the optic axis and the axis of the microscope.

Although uniaxial figures are frequently eccentric in position, the cross arms remain parallel to the planes of vibration of the nicols. Because of this fact the arms sweep the field first from

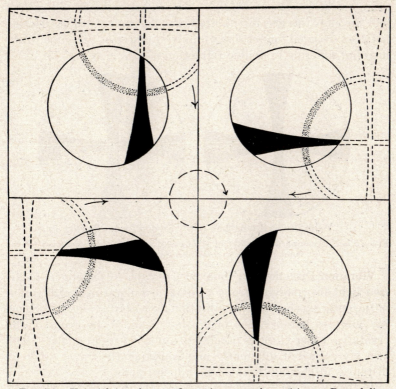

Fig. 74.—Uniaxial interference figure in eccentric positions. Dotted lines indicate the movement of the figure around the field of the microscope as the stage is rotated.

one side, then from another as the stage is rotated. It is important to note whether the arms remain parallel to the crosshairs, since arms in certain biaxial figures also cross the field. The latter are curved or crescent shaped, however, and swing across the field rather than sweep parallel to the nicols. Several eccentric positions of a uniaxial figure are shown in Fig. 74.

The number of color bands in uniaxial interference figures varies with the thickness of the section and the double refraction of the mineral. For example, thick sections of a uniaxial mineral may give a number of orders of colors, whereas a thin section of the same mineral may not yield bands of color above the first order. On the other hand, if two plates are made of different minerals, both of identical orientation and having the same thickness, the mineral with the greater double refraction will develop the greater number of color bands. The relation between uniaxial figures due to mineral plates of the same thickness but differing in double refraction is shown in Fig. 75.

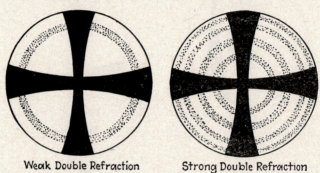

Weak Double Refraction   Strong Double Refraction

FIG. 75.—The comparative effect of strong and weak double refraction on the color bands of a uniaxial interference figure.

Vibration Directions in Uniaxial Crystals.—Convergent polarized light on emerging from a uniaxial mineral in the direction of the optic axis has specific vibration directions. One significant ray vibrates parallel to a plane that includes the *c*-axis of the crystal; another vibrates parallel to a plane at right angles. The two are refracted differently and consequently travel different distances in passing through the mineral plate.

In the upper nicol, resolution occurs into the plane of vibration of the nicol. When the rays vibrate parallel to the nicols, resolution is zero, and darkness occurs—hence the axial cross. At the 45° position of stage rotation the greatest intensity occurs, and the interference colors are most brilliant.

When two sets of rays are formed by the passage of light through a uniaxial crystal, one set travels with uniform velocity in all directions and is known as the *ordinary ray;* the other varies in velocity with direction and is called the *extraordinary ray.* If light were to radiate out from the center of a solid mass of such an anisotropic medium, at a given instant the wave front of the ordinary ray would be spherical, whereas the wave front of the extraordinary ray would be ellipsoidal. Any section of the wave front produced by the ordinary ray would therefore be

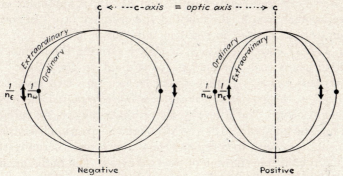

FIG. 76.—Sections of ray surfaces for uniaxial minerals.

a circle. One section of the wave front due to the extraordinary ray would be a circle; the others would be ellipses. Figure 76 illustrates significant sections. When the velocity of the extraordinary ray is greater than that of the ordinary ray, the ellipse lies outside the circle, and the mineral is optically negative. When the velocity of the ordinary ray is greater than the velocity of the extraordinary ray, the ellipse lies within the circle, and the mineral is optically positive.

The velocities represented in the diagram Fig. 76 are the reciprocals of the indices of refraction. The ray velocities have equal values in the direction of the c-axis, where the circle and ellipse coincide, and have their maximum difference

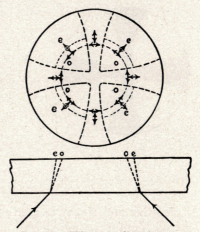

FIG. 77.—Vibration directions in a uniaxial positive interference figure. o = fast ray (least refracted); e = slow ray (most refracted). Velocity of $o = \dfrac{1}{n\omega}$; velocity of $e = \dfrac{1}{n\epsilon}$.

in a direction at right angles to the c-axis. The greatest and least indices of refraction occur at right angles to the c-axis, and in these directions (only) the indices of refraction are the reciprocals of the ray velocities.

The indices of refraction of the two rays at right angles to the *c*-axis are represented by n_ϵ and n_ω. n_ϵ is the index of the extraordinary ray, n_ω the index of the ordinary ray. In positive minerals n_ϵ is greater; in negative minerals n_ϵ is less.

In Fig. 77 convergent light is shown striking the surface of a mineral plate such as quartz, cut normal to the *c*-axis. The convergent beam is refracted and broken into two rays. One of the rays, the extraordinary ray *e*, is more refracted and has the lesser velocity. The other ray, the ordinary ray *o*, is less refracted and has the greater velocity. Although the diagram is simplified by using two lines to represent the *e* and *o* rays, actually there are many multiples of each of the two rays. The radial arrangement, however, obtains throughout.

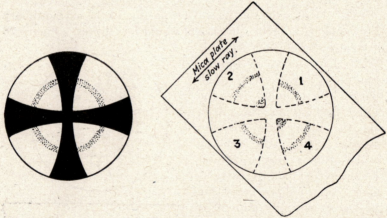

Fig. 78.—Determination of the optic character or sign for a uniaxial positive mineral.

Positive and Negative Sign of Uniaxial Crystals.—As already stated, doubly refracted rays of the uniaxial interference figure are arranged radially as shown in Fig. 77. The extraordinary ray vibrates in the *principal plane* parallel to the *c*-axis; the other vibrates at right angles. In some minerals the ray vibrating in the principal plane is the slow ray of the crystal; in others it is fast. If it is the slow ray, the mineral is positive; if fast, it is negative. The mineral in the case of Fig. 77 would be optically positive since the slow ray *e* vibrates parallel to the *c*-axis.

The position of the slow ray with reference to the *c*-axis may be determined with an accessory plate. If a mica plate, gypsum plate, or quartz wedge is inserted with the slow ray in coincidence with the slow ray of the interference figure, the color bands will change position sufficiently to indicate the optical character of the figure. If the retardation is increased parallel to the slow ray of the interference figure, the mineral is positive. If decreased, the mineral is negative. The movement of the color bands, showing the increase and decrease in retardation when a mica plate is inserted, is illustrated in Fig. 78. The color bands in quadrants 1 and 3 move toward the center; the corresponding color bands in quadrants 2 and 4 move away from the center. The movement in quadrants 1 and 3 represents increase in retardation, whereas that in quadrants 2 and 4 represents decrease in retardation. In the illustration retardation increases parallel to the slow ray since the vibration direction of the slow ray of the mica plate is parallel to quadrants 1 and 3.

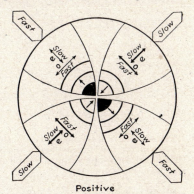

Positive

Fig. 79.—The vibration directions in both accessory plate and mineral for a uniaxial positive figure.

Examination of Fig. 79 will show that in the four parts of the circle at 45° to the planes of the nicols the extraordinary and ordinary rays lie in 45° planes or normal to 45° planes. The arrangement is also alternate and opposite.

The direction of vibration of the slow ray should be marked on each accessory. If a mica plate is inserted with the slow ray in the (1–3) position, the retardation along the extraordinary ray in the (1–3) quadrants will in effect be reinforced. At the same time, an effect of subtraction will occur in the (2–4) quadrants. The color bands of the interference figure will be displaced by this superposition. Where reinforcement occurs, the bands will move toward the center of the circle. Where subtraction occurs, the bands will move in the opposite direction.

In optically positive minerals subtraction occurs at right angles to the direction of the slow ray in the accessory. In negative minerals the subtraction is in the quadrants lying along the slow-ray direction.

The significant directions are shown in Fig. 79. The diagram indicates the direction of vibration for each of the four quadrants of a uniaxial positive interference figure in the 45° position. Corresponding slow- and fast-ray vibration directions for an accessory plate are indicated along the margins of the interference figure. The ordinary ray *o* is less refracted in the mineral and travels with greatest velocity. The extraordinary ray *e* is more refracted and travels with the least velocity.

In uniaxial negative minerals the situation is reversed. The extraordinary ray will be the fast ray, and the ordinary ray

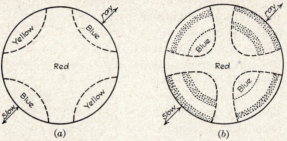

Fig. 80.—(*a*). Uniaxial positive. Quartz cut perpendicular to the optic axis as viewed in the interference figure with a gypsum plate. (*b*) Uniaxial negative. Calcite cut perpendicular to the optic axis as viewed in an interference figure with a gypsum plate.

will be the slow one. The radial arrangement of vibration directions, however, will remain the same. As a result, increase in retardation will occur parallel to the slow ray. When a mica plate is inserted, decrease in retardation produces two black dots in alternate quadrants at the center of an interference figure. The direction of the two dots forms a plus with the vibration direction of the slow ray of the mica plate in positive uniaxial minerals and a minus when the minerals are negative. This relationship serves to keep in mind the fast- and slow-ray vibration directions in uniaxial crystals.

The gypsum plate is frequently more useful for determining the optical character of a uniaxial mineral than is the mica plate. Two bright blue areas form in opposite quadrants of the interference figures of many uniaxial minerals. These stand out particularly in figures given by minerals of moderate or intermediate double refraction. When the optical character is positive, as in the case of quartz, the two blue areas occur in opposite quadrants parallel to the slow-ray vibration direction of the gypsum plate (see Fig. 80a). When the optical character is negative, as in the case of calcite, the two blue areas occur in opposite quadrants at right angles to the slow-ray vibration direction of the gypsum plate (see Fig. 80b).

Biaxial Interference Figures. *Introduction.*—Under normal conditions minerals crystallizing in the orthorhombic, monoclinic, and triclinic crystal systems give biaxial interference figures. Rarely, because of crystallization under strain, hexagonal or tetragonal minerals, normally uniaxial, are anomalous and produce biaxial figures. The latter, however, are exceptions to the general rule.

Double refraction, orientation, and thickness of section govern the character of biaxial interference figures as rigidly as in the case of uniaxial interference figures. Biaxial interference figures are also produced by the same optical arrangement of the microscope employed in the case of uniaxial figures. Unlike uniaxial figures, curves of biaxial figures assume different relative forms as the stage is rotated.

Figure 73 illustrates the two different forms of a biaxial interference figure given by a mineral at 90° and 45° intervals of rotation of the microscope stage. The 45° position is the most useful for ordinary optical determinations and is ordinarily employed in the study of biaxial minerals. The figure in this position is described as an *acute bisectrix* figure at 45°.

The 45° Acute Bisectrix Figure.—Figure 81 indicates the nomenclature of the parts of an acute bisectrix figure at 45°. The different features may be described as follows:

Isogyres.—The two broad black curves, or brushes, which mark the areas of extinction, are known as isogyres. Strong dispersion produces red and blue fringes on the margins of the isogyres. By noting the distribution of the colored fringes in the interference figure one determines the character of the dispersion. In minerals with strong dispersion the curves are not so black or so sharp as in the case of minerals with weak dispersion.

Points of Emergence of the Optic Axes.—The vertices of the two crescentlike curves mark the points of emergence of the optic axes. The amount of separation of these points differs with different minerals but is a constant for an individual mineral. The line between the two points of emergence subtends the optic axial angle.

Johannsen has suggested the word *melatope* for the points of emergence.

Plane of the Optic Axes.—The plane of the optic axes, or axial plane, includes the two points of emergence of the optic axes, the acute bisectrix direction, and the obtuse bisectrix direction.

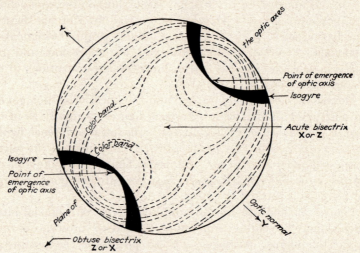

Fig. 81.—The parts of a biaxial interference figure perpendicular to the acute bisectrix in the 45° position.

Color Bands.—Interference color bands representing positions of equal retardation are distributed in symmetrical curves around the points of emergence of the optic axes and are called *isochromatic curves.*

X, Y, and Z: The three axes X, Y, and Z are distributed in the interference figure as shown in the diagram. Y is normal to the plane of the optic axes. If the acute bisectrix is X, the obtuse bisectrix is Z, and vice versa.

Optic Normal.—The direction at right angles to the plane of the optic axes is referred to as the optic normal. It is the axis Y.

Eccentric Biaxial Figures.—Since biaxial minerals as observed in thin section may be cut at any angle, a variety of modifications of the biaxial interference figures result. A single isogyre may swing across the field in one figure, another may yield an optic axis, another may show the acute bisectrix, etc. The most useful figures for optical determinations of mineral properties are either acute bisectrix or optic-axis figures. In optic-axis figures (see Fig. 93) the convex side of the isogyre in the 45° position indicates the direction of the acute bisectrix.

Optic-axis figures and most acute bisectrix figures are given by mineral sections showing comparatively low-order colors between crossed nicols in parallel light. Examination of a number of crystals of miscellaneous orientation between crossed nicols will often quickly reveal those most likely to give interference figures of useful orientation in convergent light.

Optical Directions in Biaxial Minerals.—In all biaxial minerals the various optical features may be conveniently oriented by reference to three axes, X, Y, and Z, arranged at right angles to each other. X, Y, and Z indicate the ease of vibration of light in the mineral. Light traveling normal to X vibrates parallel to the axis and has the maximum velocity for the mineral $1/n_\alpha$. Light traveling normal to Z vibrates parallel to the axis and has the minimum velocity for the mineral $1/n_\gamma$. The axis Y lies at right angles to the plane of X and Z. Light traveling normal to Y vibrates parallel to the axis and has an intermediate velocity $1/n_\beta$.

In a given mineral, light vibrating parallel to X will form the fast ray. Light vibrating parallel to Z is the slow ray, and light vibrating parallel to Y will be intermediate in velocity. Thus, when the direction of observation lies along the X-axis, XZ will indicate the slow ray and XY the intermediate ray; similarly, when the direction of observation is the Z-axis, ZX will be the fast ray and ZY the intermediate ray. When the direction of observation is the Y-axis, YX will be the fast ray and YZ the slow ray.

The fast- and slow-ray directions corresponding to the various directions of observation along the axes may be indicated as shown in the table on page 98.

When the direction of observation lies along the X-axis, light vibrating parallel to the plane XZ will have the greatest index of refraction, and light vibrating parallel to the plane XY will have an intermediate index of refraction. When the direction of observation lies along the Z-axis, light vibrating parallel to the plane ZX will have the least index of refraction, and light vibrat-

Direction of observation	Two rays observed	Velocities
X	Faster ray	$1/n_\beta$ = intermediate ray
	Slower ray	$1/n_\gamma$ = slowest ray
Y	Faster ray	$1/n_\alpha$ = fastest ray
	Slower ray	$1/n_\gamma$ = slowest ray
Z	Faster ray	$1/n_\alpha$ = fastest ray
	Slower ray	$1/n_\beta$ = intermediate ray

ing parallel to ZY will have an intermediate index of refraction. When the direction of observation lies along the Y-axis, light vibrating parallel to the plane YX will have the least index of refraction, and parallel to YZ will have the greatest index of refraction.

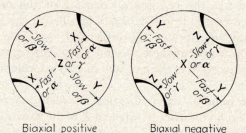

Biaxial positive   Biaxial negative

Fig. 82.—Ease-of-vibration directions X, Y, and Z, or α, β, and γ, with reference to biaxial positive and negative interference figures. Corresponding fast and slow ray directions are also indicated.

Within certain limits, the axes X, Y, and Z have positions in minerals that are dependent upon the system of crystallization. In orthorhombic minerals, X, Y, and Z are fixed with respect to the crystallographic axes a, b, and c. In the monoclinic system one of the three axes (often Y) coincides with the crystallographic axis b. In the triclinic system there are no limitations of position according to the crystallographic axes.

The optical directions in biaxial minerals may be represented in several ways. One of the most generally used devices is the ray surface illustrated in Fig. 83. Another is the index ellipsoid of Fig. 85. The ray surface is developed on X, Y, and Z arranged at right angles to each other. The index ellipsoid (optical indicatrix) may be developed on the same axes, but by convention α, β, and γ are usually used instead of X, Y, and Z. The accompanying table furnishes a comparison of the two systems of representation.

COMPARISON OF THE BIAXIAL RAY SURFACE AND THE INDEX ELLIPSOID

Comparative features	Distance from center to surface	
	Biaxial ray surface	Index ellipsoid
Axial directions		
Least velocity..............	Z	γ
Greatest velocity............	X	α
At right angles.............	Y	β
Major semi-axis..............	$1/n_\alpha$ and $1/n_\beta$	n_γ
Intermediate semi-axis.........	$1/n_\alpha$ and $1/n_\gamma$	n_β
Minor semi-axis..............	$1/n_\gamma$ and $1/n_\beta$	n_α
Optic axes..................	Secondary optic axes or biradials	Primary optic axes or binormals
Surface.....................	Double	Single

The correlation of the ease-of-vibration directions, whether designated by X, Y, and Z or α, β, and γ, with biaxial interference figures of different sign is shown in Fig. 82.

Let us assume a single crystalline mass of a biaxial crystal of sufficient size to allow examination of light variation in the system. If light were to radiate out from the center of a solid mass of such an anisotropic medium, at a given instant the wave front produced would be a double-sheeted surface with sections as illustrated in Fig. 83. The optic axes lie in the plane of X and Z and the acute angle $2V$ between the optic axes varies between 0 and 90°.

If the axis Z is the bisectrix of the acute angle between the optic axes, the mineral is said to be optically positive. If the axis X is the acute bisectrix, the mineral is said to be optically negative.

Two wave fronts appear in each section along the axes—one a circle, the other an ellipse. The size of each circle is determined by the velocity of the light ray vibrating parallel to the axis around which it is generated. Around X the radius of the circle is $1/n_\alpha$; around Y the radius is $1/n_\beta$; and around Z it is $1/n_\gamma$. Since n_α is the least index of refraction and $1/n_\alpha$ indicates the greatest velocity for the system, the circle around X is the greatest. Since $1/n_\beta$ is intermediate in velocity, the circle around Y will have intermediate size. Since $1/n_\gamma$ represents the least velocity, the circle around Z will be smaller than the circles around the two other axes. Three combinations of ellipses and circles are represented. In the section perpendicular to Y and in the plane XZ, the circle with radius $1/n_\beta$ intersects an ellipse with major and minor semi-axes $1/n_\alpha$ and $1/n_\gamma$, respectively. In the section perpendicular to Z and in the plane XY, the smallest circle, radius $1/n_\gamma$, lies within the ellipse with major and minor

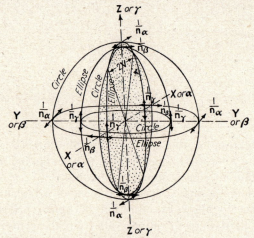

FIG. 83.—Biaxial ray surface.

semi-axes $1/n_\alpha$ and $1/n_\beta$, respectively. In the section perpendicular to X in the plane YZ the largest circle, radius $1/n_\alpha$, lies outside the ellipse with major and minor semi-axes $1/n_\beta$ and $1/n_\gamma$, respectively.

Light vibrating parallel to Z will radiate outward from the center in the plane XY. The wave front will be circular, and the velocity will be $1/n_\gamma$. Similarly, light vibrating parallel to X will travel outward in the plane YZ with a circular wave front, and the velocity will be $1/n_\alpha$. Likewise, light vibrating parallel to Y will travel in the plane XZ with a circular wave front and a velocity $1/n_\beta$. In each of these instances n_α, n_β, and n_γ represent, respectively, the least, intermediate, and greatest indices of refraction of the mineral.

The planes XY, YZ, and XZ are especially significant. Sections along each of these planes are illustrated in Fig. 84, *a*, *b*, and *c*.

In the plane XZ the ellipse and circle will cross at four points. At these four points no difference in wave velocity exists. These points of intersection mark the positions of the *secondary optic axes*, or *biradials*. In most crystals these secondary optic axes lie very near the *primary optic axes* but are not identical with them.

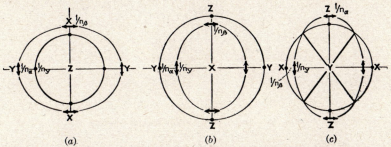

Fig. 84.—Sections of biaxial ray surface. (*a*) Section perpendicular to Z. (*b*) Section perpendicular to X. (*c*) Section perpendicular to Y.

Index Ellipsoid (Optical Indicatrix).—It is often found more convenient to represent the optical relations of crystals by means of a figure called the *index ellipsoid* (Fig. 85) than by the biaxial ray surface (Fig. 83). The index ellipsoid is also a three-dimensional figure but differs materially in development from the biaxial ray surface. The biaxial ray surface consists of two intersecting surfaces, whereas the exterior of the index ellipsoid is a single surface. The major, minor, and intermediate axes upon which the two are based also differ materially. The biaxial ray surface is based upon axes that are proportional to the reciprocals of the refractive indices. The index ellipsoid, on the other hand, is based upon axes directly proportional to the refractive indices.

The index ellipsoid or optical indicatrix for biaxial crystals may be described as a triaxial ellipsoid. In common with all triaxial ellipsoids the surface is symmetric in the origin, and in the coordinate axes and coordinate planes. The origin is the *center* of the ellipsoid, the coordinate axes are the *axes* of the ellipsoid, and coordinate planes are the *principal planes* of the ellipsoid.

The diameters of the index ellipsoid measured along the axes are $2n_\alpha$, $2n_\beta$, $2n_\gamma$. These values correspond in order to the

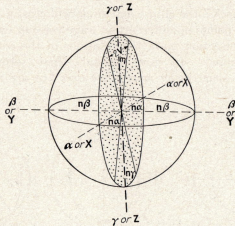

FIG. 85.—Index ellipsoid for biaxial crystals.

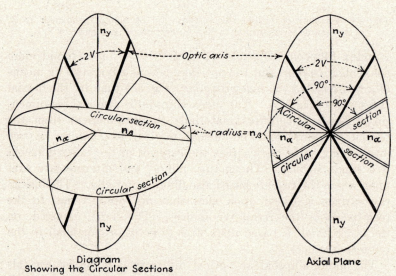

FIG. 86.—The relationship between the two circular sections, the optic axes, and ellipsoidal axes (n_α, n_β, and n_γ) in the index ellipsoid.

minor, intermediate, and major axes of the ellipsoid. The sections cut by the principal planes are the *principal sections* of the ellipsoid. These sections are ellipses and have as major and minor diameters combinations of $2n_\alpha$, $2n_\beta$, and $2n_\gamma$.

All except two of the plane sections of the index ellipsoid that are cut through the center are ellipses. These are circles (Fig. 86). The two circular sections include the semi-axis with length n_β, and thus the length of the radius of each circular section equals n_β. The directions perpendicular to the two circular sections are called the *optic axes*, or *binormals*. These are sometimes called the *primary optic axes* and differ slightly from the secondary optic axes (biradials) of the biaxial wave surface.

The index ellipsoid for uniaxial crystals provides a special case. In such crystals two of the diameters have the same value, $2n_\omega$. As a result, the principal section containing the two diameters, each having the value $2n_\omega$, is a circle. The surface of this indicatrix is an ellipsoid of revolution.

The index ellipsoid for isometric crystals is another special case. The three diameters have equal values, $2n$, and the surface developed is a sphere.

The optical properties of light rays may be determined in any given direction in a triaxial ellipsoid. Let us suppose Fig. 87 to represent the ellipsoid. The semi-axes are n_γ, n_β, and n_α, respectively, and $S'S$ represents the direction of propagation of light along a given line. If the direction of $S'S$ has been determined or is known, the following three pairs of optical properties become known by construction:

1. The vibration directions of the two rays traveling along $S'S$.
2. The two corresponding indices of refraction, n_2 and n_1.
3. The directions of the two wave normals.

If the direction of the diameter $S'S$ is known, the position of the planes tangent to $S'S$ at the two ends of the diameter also becomes known. It is then possible to pass a parallel diametral plane through the ellipsoid intersecting the center and equidistant between the two tangent planes. The diametral plane through the center will cut an elliptical section in all but two possible positions of $S'S$. These two exceptional positions are the optic axes, and here the sections cut are circular (Fig. 87). The elliptical section furnishes measurements from which the optical properties can be determined. The diametral plane will have major and minor axes. These axes mark the vibration directions of the two rays traveling along $S'S$. The major and minor radii represent the refractive indices of the waves associated with the two rays, equaling n_2 and n_1. The wave normal corresponding to the ray propagated along $S'S$ and vibrating along the major axis lies in a plane through $S'S$ and the major axis and is normal to the axis. Similarly, the wave normal corresponding to the ray propagated along $S'S$ and vibrating along the minor axis lies in a plane through $S'S$ and the minor axis and is normal to the axis.

Drop perpendiculars from the intersection of the axes with the circumference of the ellipse (Figs. 87 and 88) upon $S'OS$; these

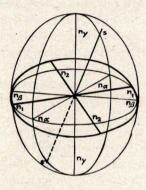

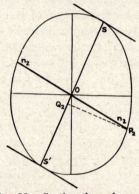

FIG. 87.—A ray OS in an index ellipsoid with a conjugate plane through O and parallel to tangent planes at S and S'.

FIG. 88.—Section through an ellipsoid showing the ray OS together with traces of tangent and diametral planes.

perpendiculars are P_2Q_2 and P_1Q_1. Then $1/P_2Q_2$ is the velocity of the ray propagated along $S'OS$ and vibrating along the major axis, and $1/P_1Q_1$ is the velocity of the ray propagated along $S'OS$ and vibrating along the minor axis. P_1Q_1 lies in a plane at right angles to the plane of the drawing in Fig. 88.[1]

The Axial Angles 2E and 2V.—The observed axial angle is always greater than the true axial angle within the mineral. This is due to the refraction of the oblique rays, as illustrated in Fig. 89. The angle $2E$ is the angle in air, while $2V$ is the internal angle.

[1] The foregoing discussion is largely based upon a paper, The Ray Surface, the Optical Indicatrix, and Their Interrelation, by Dr. George Tunell (*Wash. Acad. Sci.*, 1933).

Mallard's equation ($D = K \sin E$) may be used to determine the approximate axial angle with the microscope. In the equation K is a constant for a particular microscope, D is one-half the distance between the points of emergence, and E is one-half the axial angle in air. When $2E$ has been determined the next step is to compute $2V$.

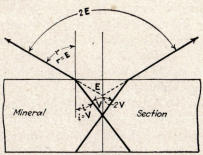

Fig. 89.—The relation between the observed angle $2E$ and the angle $2V$ in biaxial minerals.

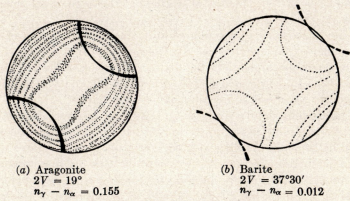

(a) Aragonite
$2V = 19°$
$n_\gamma - n_\alpha = 0.155$

(b) Barite
$2V = 37°30'$
$n_\gamma - n_\alpha = 0.012$

Fig. 90.—Comparison of axial angles.

The computation of the axial angle in a mineral from the observed axial angle in air depends upon the formula

$$\sin E = n_\beta \sin V$$

When $n \sin V$ is equal to 1, the angle $2E$ becomes 180°, and the axial angle in air cannot be measured. Angles greater than 180° likewise cannot be measured in air. The value of the observed angle may be reduced to measurable dimensions by immersing the objective in oil of known refractive index.

Large axial angles need to be measured with a rotation device. Such devices for rotating crystals in a vertical circle may be adapted to the stage of the microscope; otherwise special apparatus must be employed.

Variation in Axial Angle.—Figure 90 illustrates two biaxial interference figures in the 45° position. The figures represent two different minerals—aragonite on the left and barite on the right. The two sections from which the interference figures are derived have been cut normal to the acute bisectrix in each instance, and the sections are also of approximately the same thickness. As a result of these conditions, only two variables remain to produce differences in the diagram: variation in the axial angle $2V$ and variation in the double refraction $n_2 - n_1$.

The isogyres in the figure on the left represent the approximate position of the two curves in relation to the field of the microscope for an axial angle $2V = 19°$ (aragonite). The figure on the right represents barite drawn to the same scale. In this instance the angle $2V = 37°30'$ places the isogyres at the edge of the field of view.

The dotted lines in the figures indicate the distribution of the color bands. Aragonite has a double refraction of 0.155, which is considerably larger than the double refraction of barite, which is 0.012. In consequence, for the same thickness of section, aragonite has many more color bands than barite.

The student should study the interference figures of a number of different mineral sections cut normal to the acute bisectrix until he becomes familiar with the variation of the isogyres with the axial angle. In fact, it is worth while to record in a notebook the relative positions of the isogyres for angles in the neighborhood of 5, 10, 15, 20, 25, 30, 35, and 40°. A record of this sort will be of considerable assistance in determining the approximate axial angle of an unknown mineral.

It should also be remembered that if the thickness remains the same, the number of the color bands of the interference figures will either increase or decrease with increase or decrease in the double refraction.

Determination of the Optic Sign of a Biaxial Mineral.—The optic sign is best determined with the mineral in the 45° position. The quartz wedge is employed for most determinations. In some cases, however, a mica plate or gypsum plate may be preferred. The same principles utilized when the quartz wedge is employed apply equally in determinations with the other accessory plates.

As stated before, X, Y, and Z are the axes of ease of vibration. Light traveling through a crystal normal to X has the maximum velocity for all directions in the crystal. Light traveling normal to Z has the least velocity. The direction Y is normal to the plane of X and Z. When the direction X is the acute bisectrix, the mineral is optically negative. If Z is the acute bisectrix, the mineral is positive.

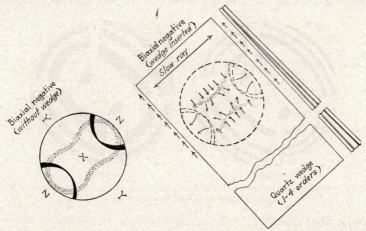

Fig. 91.—The determination of the optic sign with a biaxial negative interference figure.

A biaxial negative crystal in the acute bisectrix position at 45° will be used to illustrate the method of determining the optic sign (see Fig. 91). A biaxial figure of this type is first observed carefully in order to note the position of the color bands, both in the central area and within the two small areas inclosed by the concave portions of the isogyres. A quartz wedge is then inserted in the accessory slot with the slow ray parallel to the axial plane. Movement of the color bands occurs as the wedge is inserted.

The movement of the color bands in a negative crystal is indicated by the arrows in Fig. 91. As the wedge is moved in— *i.e.*, as the thickness increases—the color bands in the central area move toward the two "eyes," or melatopes, of the interference figure. At the same time the bands on the opposite sides of the isogyres within the two small areas move away from the melatopes. As the wedge is withdrawn, the movement of the color bands is reversed. If a positive crystal is substituted, the movement of the color bands is also reversed.

In the biaxial negative crystal illustrated the axis Z lies in the axial plane along the direction of the obtuse bisectrix. The axis X is perpendicular to Z and is the direction of the acute bisectrix. The axis Y is the optic normal. Two rays travel along X with vibration directions at right angles to each other, the vibration directions being parallel, respectively, to Z and Y. The ray vibrating parallel to Z is the slow ray for the crystal (velocity

Positive   Negative

FIG. 92.—Positive and negative biaxial crystals (indicating their appearance with a mica plate in monochromatic light).

$(1/n_\gamma)$; that parallel to Y is intermediate in velocity, having the value $1/n_\beta$. Thus, in the central area of the interference figure at X, we have a slow ray, velocity $1/n_\gamma$, and an intermediate but faster ray, velocity $1/n_\beta$. Consequently, if the slow ray of the quartz wedge is parallel to the direction Z, increase in retardation occurs as the wedge thickness increases. This results in a movement of the color bands toward the melatopes in the central portion of the figure as the wedge is inserted. At the same time the color bands in the outer portions will move in the opposite direction since here the slower ray and faster ray relations are reversed. If the quartz wedge is always inserted as indicated in Fig. 91, an acute bisectrix biaxial negative interference figure in the 45° position will always show movement of the color bands toward the melatopes in the central area. Conversely, a biaxial positive figure treated in the same way will show movement in the opposite direction. Since the slow-ray vibration direction in the quartz wedge is marked, the slow-ray vibration direction in the interference figure is easily determined by comparison. Examples of both positive and negative biaxial figures in monochromatic light with the slow-ray vibration direction of an accessory plate superimposed are shown in Fig. 92.

The Optic-axis Figure.—Interference figures produced by sections cut normal or nearly normal to one of the two optic axes of a biaxial mineral are useful for determinations of optic sign. Such sections yield interference figures having a single isogyre in the field of view. The melatope, or point of emergence, may coincide with the axis of the microscope or may be slightly off center.

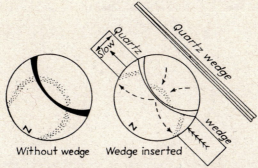

Fig. 93.—Movement of the color bands in an optic-axis biaxial positive interference figure as an accessory plate is inserted.

As the stage is rotated, the isogyre swings around the field, remaining centered or nearly centered, depending upon the eccentricity of the section. The color bands are arranged almost circularly around the melatope and vary in retardation with the double refraction of the mineral.

The curvature of the isogyre decreases with an increase in $2V$. When the axial angle is large—*i.e.*, near 90°—the isogyre is straight, and the acute bisectrix side of the interference figure becomes indistinguishable from the obtuse bisectrix side. When the angle is small, however, the isogyre is definitely curved in a crescentlike form. The convex side of the curve in the 45° position points toward the acute bisectrix, and the obtuse bisectrix is on the concave side.

Optic-axis figures showing even slight curvature are useful for determinations of the optic sign. The mica plate, gypsum plate, or quartz wedge may be employed, depending upon the double refraction of the mineral. The effect of the quartz wedge upon a biaxial positive optic-axis interference figure is shown in Fig. 93. An optic-axis figure without the wedge inserted is shown on one side of the diagram, and the movement of the color bands caused by insertion of the wedge is shown on the opposite side.

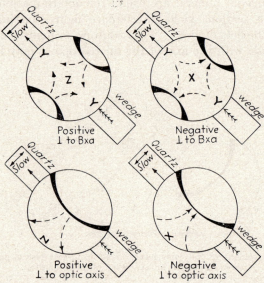

FIG. 94.—Movement of the color bands to be expected as the quartz wedge is inserted in acute bisectrix and optic-axis interference figures of opposite sign.

Diagrams are shown in Fig. 94, which should be of convenience in determining the signs of interference figures in the two most useful positions upon insertion of the quartz wedge. The same principles apply in utilizing mica and gypsum plates.

Dispersion in Biaxial Interference Figures.—The optic angle for light of one wave length is frequently greater or less than for light of another wave length. In the normal interference figure produced by white light this is usually detected by a peculiar arrangement of the color bands or by the development of blue and red fringes on the isogyres.

The dispersion recorded in most tables of optical mineralogy is that of the optic axes. The two extreme rays of the spectrum are used to designate the character of the dispersion. Thus, if the axial angle for red r is greater than that for violet v, the dispersion is expressed $r > v$. In case the reverse is true, the formula is $r < v$.

In many instances the dispersion can be determined by direct observation of the biaxial interference figure (Fig. 95). If the isogyres of the interference figure ($r > v$) have a distinct red fringe on the convex edges, the angle for red is greater than for violet. Both isogyres should be observed before reaching a

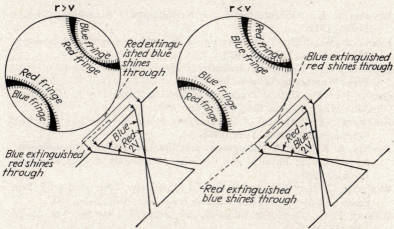

Fig. 95.—Biaxial interference figures illustrating dispersion $r > v$ and $r < v$. The colored fringes as observed in the interference figure are reversed from the axial angles existing in the crystal due to extinction.

conclusion. On the concave side of the isogyre, as illustrated in the figure, red light is extinguished, and consequently the concave fringe is blue in color. Blue is extinguished on the convex side, and the fringe is red.

It should be emphasized that the symmetry of the interference figure is always governed by the symmetry of the mineral. Dispersion varies according to the symmetry of the crystal system. The various types, arranged according to crystal system, are as follows:

> Orthorhombic crystals:
>   Dispersion of the optic axes.
>   Crossed axial plane dispersion.

Monoclinic crystals:
    Inclined dispersion (both bisectrices).
    Horizontal dispersion (acute bisectrix).
    Crossed dispersion (obtuse bisectrix).
Triclinic crystals:
    Unsymmetrical dispersion.

These types are best distinguished by the use of light of various wave lengths. Red and blue color filters to be placed in front of the mirror of the microscope are convenient.

Suggested References

BUCHWALD, E.: "Einführung in die Kristalloptik," Sammlung Göschen, Berlin, 1912.

EVANS, J. W.: "The Determination of Minerals under the Microscope," Murby & Co., London, 1928.

FLETCHER, L.: "The Optical Indicatrix," Henry Frowde, Oxford University Press Warehouse, London, 1892.

GROTH, P.: "The Optical Properties of Crystals," trans. by B. H. Jackson, John Wiley & Sons, Inc., New York, 1910.

JOHANNSEN, A.: "Manual of Petrographic Methods," McGraw-Hill Book Company, Inc., New York, 1918.

TUTTON, A. E. H.: "Crystallography and Practical Crystal Measurement," 2d ed., vol. 2: Physical and Chemical, Macmillan & Company, Ltd., London, 1922.

WINCHELL, A. N.: "Elements of Optical Mineralogy," 5th ed., Part I: Principles and Methods, John Wiley & Sons, Inc., New York, 1937.

WRIGHT, F. E.: The Index Ellipsoid, *Am. Jour. Sci.*, 4th ser., vol. 35, pp. 133–138, 1913.

————: The Formation of Interference Figures: A Study of the Phenomena Exhibited by Transparent Inactive Crystal Plates in Convergent Polarized Light, *Jour. Optical Soc. Am.*, vol. 7, pp. 779–817, 1923.

CHAPTER VII

COLOR, FORM OR AGGREGATION, CLEAVAGE, AND ORIENTATION

Color and Pleochroism.—The color of minerals in the hand specimen is not always a good index of the appearance of the mineral in a thin section. Minerals are faintly colored in thin sections in contrast to deeply colored minerals in hand specimens, and frequently minerals having definite and pronounced hues when viewed with the unaided eye are colorless beneath the microscope. Since sections that are not properly ground and that are too thick will often retain the hand-specimen color, it is particularly desirable to have thin sections of deeply colored minerals ground as thin as possible. It is obvious that deeply colored minerals are more likely to be colored in thin section than are pale minerals.

Isotropic colored minerals show no color change as the mineral is rotated in plane polarized light. Anisotropic colored minerals, on the other hand, exhibit a change in color in varying degree as the stage is rotated. Since the same mineral frequently produces a number of colors the change in color produced is known as *pleochroism.* Occasionally light vibrating in one plane through a crystal is colorless, whereas in another plane at right angles a definite hue is observed; or the condition may obtain in which two vibration directions produce colors, whereas a third may be colorless. Three different pleochroic colors may be produced by a single biaxial crystal, or two pleochroic colors by a uniaxial crystal.

Hexagonal or tetragonal colored minerals are *dichroic—i.e.,* the pleochroic coloring of minerals in these two systems as exhibited with the microscope is twofold.

Orthorhombic, monoclinic, and triclinic minerals, when colored in thin section, exhibit three different colors and are *trichroic.* The pleochroic colors are normally oriented with the axes X, Y, and Z of the crystal. When the directions of the

113

axes of ease of vibration in a mineral are known from the presence of important cleavages, crystal boundaries, twin planes, or from an interference figure, the colors of light vibrating parallel to the axes X, Y, and Z may be ascertained. Since X, Y, and Z are the vibration axes, a definite relationship exists with n_α, n_β, and n_γ and also the optic axes. The correlation of the relationships is best accomplished by study of the interference figure in order to determine the directions of the axes. As soon as the acute bisectrix Z (if the mineral is positive) or X (if the mineral is negative) is known, the Bertrand lens and analyzer should be removed from the microscope tube and the corresponding natural color ascertained. The color produced by light vibrating in a plane at right angles to the axial plane is Y. The third color will be due to light vibrating parallel to the direction of the obtuse bisectrix. This will be X if the mineral is positive or Z if the mineral is negative.

In pleochroic uniaxial minerals light vibrating parallel to the optic axis is one color, whereas light at right angles is another.

Biaxial minerals exhibit varying degrees of absorption of color. In common hornblende, for instance, light vibrating parallel to Z usually shows the most absorption, Y is less absorbed, and X is the least. This is recorded by means of the *absorption formula*

$$X < Y < Z$$

It is customary to record the absorption in terms of vibration parallel to the ease of vibration axes.

Form or Aggregation.—Minerals assume many varying forms beneath the microscope, depending upon the chemical nature of the material and the conditions of crystallization. A considerable number of the designs produced by groups of crystals observed in thin sections and cross sections of individual minerals are so unusual that the mental picture produced is of decided aid in identification.

This tendency is useful on account of the frequency with which many minerals assume a peculiar development. Such a tendency in the case of individual crystals may be described as *habit*. *Aggregation* refers to the nature of grouping of either a few or numerous small crystals. The pattern that a mineral group assumes may be described as its *mode of aggregation*. Both form and aggregation are discussed in the following paragraphs.

Incipient Crystallization.—Natural glass frequently forms from a viscous liquor that upon solidifying lacks crystallization and is isotropic. The material, however, contains constituents capable of producing a number of different minerals. Development of these minerals is hindered by the viscosity of the inclosing liquor during the period of crystallization, largely because of quick release of pressure and rapid cooling of the material. As a result, definite crystals may not develop, but instead needlelike aggregates, fernlike growths, and various odd designs may be formed representing the sudden arrest of the process of crystallization. *Crystallites, margarites, trichites, microlites, globulites,* and *longulites* are names that have been applied to various forms of incipient crystallization. Trichites are curved streaks of embryonic crystals in glass. Margarites are long streaks of globular forms resembling portions of strings of seed pearls in curved or straight lines. Longulites are small rodlike forms composed of the grouping of globulites. Crystallites are the minute nuclei of crystallization existing suspended in volcanic glass. Microlites are small needlelike, almost crystalline forms. Incipient crystals in glass are illustrated in Fig. 96.

Non-crystalline Isotropic Minerals.—Minerals lacking directional qualities capable of producing double refraction are dark between crossed nicols. Such minerals are chiefly identified by means of their structure in thin section, combined with a determination of their indices of refraction.

Glass, opal, cliachite (bauxite), and collophane are illustrations of common non-crystalline or amorphous minerals.[1] In addition to incipient crystals, glass frequently exhibits flow lines, cracks, or concentric fractures of distinctive character. Opal is usually banded and in addition may exhibit a play of colors. Shatter cracks are distinctive features in minerals of colloidal origin (Fig. 97).

Cliachite, the amorphous mineral making up a considerable portion of bauxite, occurs in pisolitic or rounded forms frequently cracked and fractured at random. The interstices between the pisolitic forms may be filled with finely crystalline gibbsite forming a fine-grained gray-and-white mosaic of small crystals between crossed nicols or may be transparent without crossed

[1] The term *amorphous* is used here, although X-ray studies have shown these materials to have directional properties sensitive to short wave lengths.

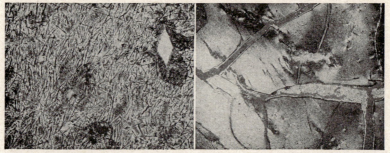

Fig. 96.—Incipient crystals of rod-like aggregates in a groundmass of volcanic glass.

Fig. 97.—Shatter cracks in halloysite (crossed nicols).

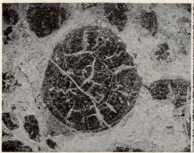

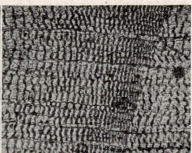

Fig. 98.—Rounded sections of pisolitic bauxite made up of the mineral cliachite. Interstices (around holes in the section) with gibbsite.

Fig. 99.—Cellular structure of wood preserved in opal.

Fig. 100.—Chalcedony between crossed nicols showing both radial aggregates and banded structure.

Fig. 101.—A flamboyant radial aggregate of sillimanite between crossed nicols.

nicols. A thin section of cliachite and gibbsite in bauxite rock from Arkansas is illustrated in Fig. 98.

The rounded, more or less spherical forms assumed by amorphous and metacolloidal minerals in open spaces are frequently described as *colloform.*

Materials of Organic Origin.—In the study of thin sections, structures are occasionally encountered that are residual from organic life. Diatomaceous, radiolarian, or foraminiferal organisms have distinctive structures frequently preserved in minerals. Fossil diatoms and other microscopic remains are usually opal and have the optical properties of that mineral. The original structures of the microorganisms, however, are on many occasions well preserved (Fig. 102). Foraminifera are apt to yield calcite usually finely crystalline in nature and difficult to distinguish from aragonite unless the structure is sufficiently coarse to permit the rhombohedral cleavage to develop (Fig. 103).

Fragments of former vegetable matter preserved as carbonaceous material—lignite, etc.—are usually black or brown in thin section. Cellular structures of wood as preserved in lignite are quite distinctive in thin section. Opal formed by the replacement of wood frequently exhibits cellular structure (Fig. 99).

Collophane, the mineral constituent of fossil bone, usually retains the structure of the bone that has been replaced (Fig. 196).

Fine Aggregates.—Minerals frequently form fine aggregates of distinctive pattern. Aggregate structure is emphasized between crossed nicols either by radial groups or by a fine-grained mosaic-like groundmass of small crystals.

The radial structures and also fine-grained mosaics (Fig. 104) furnish excellent illustrations of this feature. In radial groups the small crystals converge like the spokes of a wheel. Radial uniformity of orientation produces with crossed nicols a dark cross parallel to the positions of extinction. This cross should not be confused with the axial cross of a uniaxial interference figure produced by conoscopic observation of a single crystal and having an entirely different significance.

An illustration of a mineral aggregate is furnished at times by sillimanite (Fig. 101). The needles of sillimanite, however, are not so regularly arranged as is the case of chalcedony.

Crosses due to radial arrangement of fine crystal groups in polarized light are frequently formed in microfossils replaced by

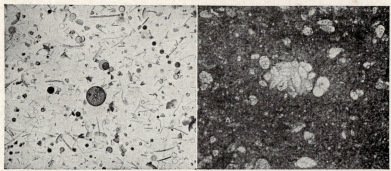

FIG. 102.—The fine microorganic structures preserved in opal-forming fossil diatoms. Diatomaceous earth from Lompoc, California.

FIG. 103.—Calcite in sections of fossil foraminifera scattered through carbonaceous shale.

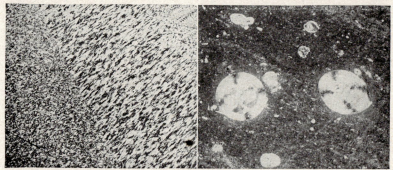

FIG. 104.—Mosaic structure in chalcedony. A "salt and pepper" aggregate of small crystals photographed between crossed nicols.

FIG. 105.—A polarization cross in fossil foraminifera of calcite arranged in concentric bands of radial fiberlike crystals.

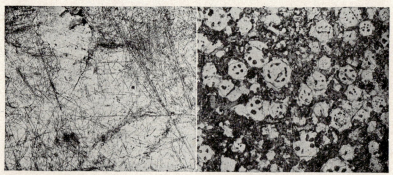

FIG. 106.—Needlelike crystals of sillimanite in quartz.

FIG. 107.—Euhedral crystals of leucite containing inclusions of glass.

calcite. Figure 105 illustrates fossil foraminifera in shale, the outer portions of the two large structures showing parts of an aggregate polarization cross.

Spherulitic crystallization in glassy flow rocks is strikingly illustrated between crossed nicols. Figure 148 represents a thin section containing spherulites photographed between crossed nicols. Spherulites of the feldspars or other minerals are often found suspended in volcanic glass of one form or another.

Parallel orientation of fibrous aggregates is illustrated by a thin section of chrysotile in serpentine (Fig. 108). Veinlets of cross-fibers arranged in parallel fashion perpendicular to the walls of the vein are common in serpentine. The fibers are moderately anisotropic, whereas the serpentine is almost dark between crossed nicols.

Inclusions.—At many times during the process of crystallization small areas of foreign substances are caught within what are otherwise clear crystals. In the crystallization of leucite, for example, small areas of volcanic glass are often distributed symmetrically as small isolated spheres suspended in the crystal. Leucite of this type is illustrated in Fig. 107.

Hypersthene (Fig. 110) also contains areas of brown, flakelike inclusions frequently accompanied by a fine transverse system of lines. This is usually described as *schiller structure*.

Occasionally the substances retained within the crystal during formation may be radioactive and during the long progress of geologic time will continue to give off emanations until they finally lose their strength. Such inclusions, when trapped in colored minerals such as biotite, produce dark brown circular patches, frequently pleochroic. Figure 111 illustrates halos produced by radioactivity in biotite from western Connecticut.

Needlelike Crystals.—A few minerals form fine, hairlike masses of crystals, usually penetrating some other mineral, such as mica or quartz. Sillimanite, for example, may be found in minute needles penetrating quartz in all directions (Fig. 106). Dumortierite occurs in a similar manner. The dumortierite, however, is usually pink and may impart to the quartz in the hand specimen a color resembling rose quartz, although deeper in color. Rutile forms red or brown needles that may penetrate either quartz or mica. Tourmaline may also occur in similar fashion. The

FIG. 108.—Bands of asbestos fibers in serpentine (crossed nicols).

FIG. 109.—Needlelike crystals of tourmaline arranged in radial groups.

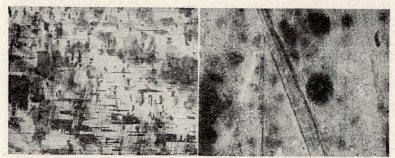

FIG. 110.—Schiller structure in hypersthene.

FIG. 111.—Pleochroic halos in biotite.

FIG. 112.—Bladed crystals of kyanite.

FIG. 113.—Sphene twinned parallel to the length of the section.

radiating crystals of tourmaline in quartz illustrated in Fig. 109 are characteristic of luxullianite, a rock.

Although these occurrences are quite striking when observed, it should be remembered that the same minerals may occur in large crystals having an entirely different habit.

Bladed Crystals.—Crystal groups may be composed of larger, coarser individuals causing lathlike sections under the microscope. Also, intermediate sizes of different form and development may occur. One illustration of a coarse-bladed type of development is furnished by kyanite, as illustrated in Fig. 112.

Twin Crystals.—The feldspars provide the most outstanding illustration of twinning. Twin lamellae stand out particularly between crossed nicols. The twinning is for the most part polysynthetic, comprising multitudinous lathlike individuals oriented according to either the albite or the pericline law or perhaps both (Fig. 216). Orthoclase provides good illustrations of twinning according to the Carlsbad law and occasionally according to the Baveno and Manebach laws. Plagioclase also twins according to the Carlsbad law. A most outstanding illustration of twinning is the grid structure of microcline between crossed nicols as shown in Fig. 209.

Other minerals besides feldspars provide good illustrations of twinning. Calcite nearly always twins parallel to the long diagonal of the cleavage rhombohedron, and dolomite twins parallel to both the long and the short diagonal of the cleavage rhombohedron. Cassiterite, corundum, pyroxene, aragonite, amphibole, lazulite, and gibbsite are frequently found in twin crystals, the twinning being easily recognized by the different extinction of the various twin individuals between crossed nicols. The crystal of sphene in Fig. 113 is twinned into individuals separated by a plane parallel to the length of the crystal.

Natural Crystal Form in Thin Section.—There is a pronounced tendency among crystallized minerals of a given species to follow the same habit of growth. In such instances outlines viewed in thin section, due to natural crystal form, are significant. These outlines often suggest the identity of a mineral at a glance. Apatite, for example, frequently appears in small, lathlike, elongated crystals with hexagonal cross sections (Fig. 117).

Corundum in mica schist may form skeleton crystals characterized by rounded elongate outlines (Fig. 118).

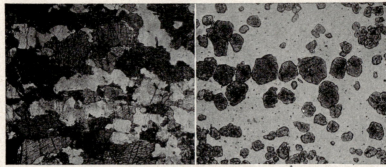

FIG. 114.—Anhedral crystals of hornblende associated with quartz.

FIG. 115.—Subhedral garnet in quartz.

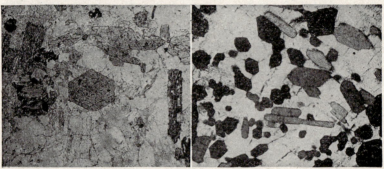

FIG. 116.—Euhedral hornblende associated with biotite in thin section.

FIG. 117.—Apatite crystals in thin
section (crossed nicols).

FIG. 118.—Skeleton crystals of corundum in mica schist.

FIG. 119.—Andalusite containing
symmetrically arranged carbonaceous
inclusions.

Pyrite is often found in square areas in thin sections. Although triangular and other shapes also occur, opaque sections of cubes having a brass-yellow color in reflected light distinctly identify the mineral.

Crystals of the type just described occurring in thin sections with well-developed crystal boundaries are called *euhedral*.[1] *Anhedral*[2] crystals are crystals with rounded or irregular boundaries. Partially developed crystals may be called *subhedral*.

Anhedral, subhedral, and euhedral crystals are illustrated in Figs. 114, 115, and 116, respectively. Subhedral garnet (Fig. 115) should be compared with euhedral garnet (Fig. 283).

Isometric Crystals.—Euhedral crystals of common isometric minerals encountered in thin sections frequently exhibit cross sections of such common isometric forms as the cube, octahedron, dodecahedron, and trapezohedron.

Leucite and analcime are illustrated in Figs. 107 and 120. The outline of the analcime crystal and also of the leucite crystals follows the trapezohedron.

Occasionally isometric crystals are twinned, and in some cases weak anisotropism exists, although isometric minerals are normally isotropic. Pseudo-isometric minerals such as leucite exhibit low first-order interference colors. Garnet from contact-metamorphic deposits in limestone may be strongly anisotropic at times. An illustration of garnet of this type is given in Fig. 121.

Tetragonal Crystals.—Few tetragonal minerals having euhedral development are encountered in the routine examination of thin sections. Cross sections of tetragonal minerals are usually the normal sections to be expected from a tetragonal prism terminated with a bipyramid, as in the case of zircon, or a combination of two prisms and two bipyramids, as in the case of idocrase. The crystals are uniaxial, and sections cut normal to the *c*-axis give optic-axis interference figures. Twinning and cleavage are arranged in accord with tetragonal symmetry. Crystals are anisotropic and extinguish parallel to the crystallographic axes.

Hexagonal Crystals.—A number of common minerals encountered in thin section are hexagonal in crystallization. The

[1] Also called *idiomorphic* or *automorphic*.
[2] Also called *allotriomorphic* or *xenomorphic*.

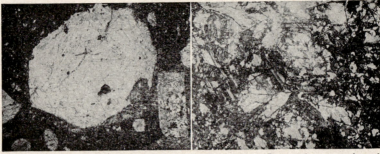

FIG. 120.—A euhedral crystal of analcime.

FIG. 121.—Isometric crystals of garnet showing anomalous anisotropism and banding between crossed nicols.

FIG. 122.—Nepheline crystals showing both hexagonal and rectangular outlines.

FIG. 123.—A euhedral crystal of olivine.

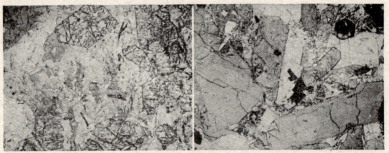

FIG. 124.—Euhedral staurolite crystals in a groundmass of muscovite and other minerals.

FIG. 125.—Euhedral crystals of pyroxene.

crystals follow two principal types of development: either the crystallization is in accord with the hexagonal subsystem, or the rhombohedral subsystem is favored. The hexagonal subsystem is usually represented by elongated crystals of hexagonal cross section, whereas sections of the rhombohedral division are nearly always rhombic and only occasionally hexagonal.

Quartz, tourmaline, apatite, beryl, and nepheline are frequently encountered hexagonal crystals. The sections of these minerals may be hexagonal or triangular if normal to the *c*-axis, and rectangular or modified rectangular where cut along the *c*-axis. The crystals are commonly elongated parallel to the *c*-axis and exhibit parallel extinction with positive or negative elongation, depending upon the sign of the mineral. The hexagonal sections are either nearly or completely isotropic and yield uniaxial interference figures. Nepheline (Fig. 122) frequently exhibits both hexagonal and rectangular sections. Apatite (Fig. 117) furnishes both hexagonal and elongated sections.

Rhombohedral crystals include the group of rhombohedral carbonates, and although these minerals are ordinarily found in matted masses of anhedral crystals, occasionally euhedral crystals occur. The euhedral crystals are rhombic in section and frequently exhibit cleavage lines parallel to the sides. The section of the rhombohedron normal to the *c*-axis may appear hexagonal.

Orthorhombic Crystals.—A number of orthorhombic minerals encountered in thin section form euhedral crystals. Olivine, natrolite, barite, zoisite, andalusite, dumortierite, and topaz are among the most common. Euhedral olivine crystals (Fig. 123) are frequently seen in thin sections of basic igneous rocks. The crystals are symmetrical with respect to the crystallographic axes and become extinct when the axes are parallel to the plane of vibration of the nicols. Long square crystals of natrolite with excellent parallel extinction mentioned in the discussion of the adjustments of the microscope are occasionally found in thin sections. Barite crystals may be square, rectangular, or elongated, with parallel or symmetrical extinction.

Staurolite crystals are illustrated in Fig. 124. Quartz inclusions are common in crystals of this type. Andalusite contains carbonaceous inclusions arranged in a symmetrical rhombic pattern (Fig. 119).

Monoclinic Crystals.—The pyroxenes, amphiboles, monoclinic feldspars, sphene, mica, epidote, and a number of other less common monoclinic minerals are frequently found in euhedral crystals in thin sections.   The crystals exhibit inclined extinction when sections are cut either parallel to or near the plane of the *a*- and *c*-axes.   Certain sections, however, are frequently so oriented as to furnish either symmetrical or parallel extinction. More crystals are necessary for a proper identification and study of monoclinic minerals in thin section than are needed for minerals in the other systems just described.   Each monoclinic crystal is an individual problem in optical orientation and should be considered by itself.

Figures 116 and 125 furnish comparative examples of euhedral amphibole and euhedral pyroxene.   The views demonstrate both crystal boundaries and cleavage.

Triclinic Crystals.—The plagioclase group, microcline, kyanite, and rhodonite constitute the ordinary triclinic minerals found in thin section.   The extinction is ordinarily inclined in all sections, and the extinction angles are frequently high.   The plagioclase group and microcline exhibit striking twinning in addition, the nature of the twinning helping considerably in differentiation.

Cleavage, Parting, and Fracture as an Aid in Distinguishing Minerals.—Cleavage may be defined as the ever-present ability of a mineral to break into smaller and smaller particles, with smooth surfaces along planes parallel to the directions of the faces of certain possible forms.   Cleavage is frequently of assistance in distinguishing minerals.   Unfortunately, many minerals show little or no cleavage.   When cleavage is well developed, however, a mineral may be often easily identified by means of this property.   In the grinding of thin sections, cleavage planes are usually developed and appear in the finished section as lines of varying width.

Some minerals separate only occasionally or break along planes of twinning.   Since this ability to separate is not always present and may not continue to finer and finer particles, it is usually described as *parting*.   In an individual specimen, in so far as the effect produced is concerned, cleavage may be indistinguishable from parting.

Several types of cleavage may be observed in minerals. These are often seen in the examination of both thin sections and mounts of crushed fragments. The same type of cleavage often appears considerably different in fragments than it would appear in thin section. This is due to the fact that fragments normally fall with cleavage surfaces flat or almost flat on the surface of the slide, whereas in thin sections a large number of random orientations are rigidly fixed in the plane of the section.

In the appended tables for identifying common minerals encountered in the study of thin sections, cleavage planes, fracture planes, or the tendency to break parallel to certain definite directions are indicated for each of the minerals included.

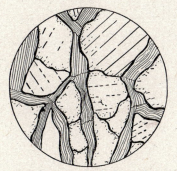 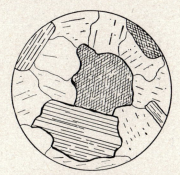

FIG. 126.—Anhedral t o p a z crystals with broadly spaced cleavage in one direction with penetrating muscovite having closely spaced cleavage in one direction.

FIG. 127.—Anhedral hornblende with cleavage at 56° and 124° in a groundmass of feldspar.

Cleavage in One Direction.—A considerable number of minerals have a single plane of cleavage. Muscovite and topaz are examples (Fig. 126). In thin section, crystals showing one direction of cleavage usually exhibit systems of parallel lines, the lines being either closely spaced or spaced at considerable distances. Occasionally, a cleavage plane may be almost parallel to the section, in which instance practically no cleavage lines will appear.

In the case of fragments, minerals with one direction of cleavage usually lie flat upon the microscope slide and have irregular boundaries. The interference color is nearly always uniform for the area of the fragment except on the outer edge, where a number of color bands will be observed, increasing in number in the case of minerals having higher double refraction.   Frequently similar orientation of interference figures occurs in the fragments since many fragments lie in the same position.

Cleavage in Two Directions.—Several common minerals develop prominent cleavage in two directions.   The pyroxenes, amphiboles, and feldspars are outstanding illustrations.

Common hornblende of the amphibole group is distinguished from pyroxene by a difference in cleavage angle.   The cleavage of hornblende (Fig. 127) parallel to the rhombic prism {110} is in two directions at 56 and 124°.   The cleavage of the rhombic

Fig. 128.—Euhedral pyroxene crystals showing two directions of cleovage in a groundmass of feldspar.

Fig. 129.—Andalusite with cleavage in two directions at about 90°, cut by veinlets of fine muscovite with cleavage in one direction.

prism in the case of pyroxene (Fig. 128), however, is approximately 87 and 93°.

The two directions of cleavage may be at 90° in orthorhombic or tetragonal minerals.   Figure 129 illustrates andalusite (orthorhombic), with two directions of cleavage at about 90° (89°12').

Cleavage in Three Directions.—The types of cleavage produced due to cleavage in three directions vary considerably.   One of the simplest types is that produced by cleavage parallel to the faces of the cube.   In thin sections cleavage of this type produces square or triangular patterns.   In fragments the boundaries tend to be square or rectangular.   Cubic cleavage is restricted to the isometric system, and both sections and fragments are easily confirmed by the isotropic character of the material.

Rectangular cleavage is similar to cubic cleavage in its appearance in both thin sections and fragments (Fig. 130). Minerals having rectangular cleavage, however, may be easily distinguished from those with cubic cleavage, owing to their anisotropism between crossed nicols.

Rhombohedral cleavage is one of the most common types found within minerals, both in thin sections and in fragments. The common carbonate minerals have cleavage of this type. The pattern produced in thin section is crisscross in design, and the crystals usually show inclined cleavage planes penetrating the section. In addition, there is frequently a set of twin lines parallel to the long diagonal of the rhombohedron. Occasionally twinning may appear parallel to both the long and the short diagonal, as in the case of dolomite. In fragments, minerals with rhombohedral cleavage are usually flat lying, with a fairly well-developed rhombic section, and vary in relief with direction. The edges usually show wedgelike or inclined surface boundaries.

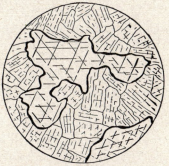

Fig. 130.—Anhydrite (rectangular cleavage) and fluorite (octahedral cleavage) in the same thin section. The octahedral cleavage of fluorite produces a triangular pattern.

Cleavage in three directions is frequently produced by breaking parallel to various directions in crystals of the orthorhombic, monoclinic, and triclinic systems. In most instances, the part of the mineral under examination presents a special case. As a rule, however, cleavage of this type produces a crisscross of almost rectangular pattern in thin section, and in fragments the boundaries either are almost rectangular or may be flat lying, with wedgelike edges.

Cleavage in Four Directions.—One common mineral, fluorite, has cleavage in four directions parallel to the faces of an octahedron (Fig. 130). Cleavages of fluorite in thin section tend to develop triangular or rhombic patterns. In fragments the outlines of individual fragments are triangular or irregular with pointed edges. These fragments are easily detected between crossed nicols on account of the isotropic character of the mineral. Occasionally spinel is found with an octahedral parting imperfectly developed but somewhat resembling octahedral cleavage. The diamond has octahedral cleavage but, needless to say, is not retained in grinding thin sections.

Cleavage in Six Directions.—Sphalerite is one of the few minerals with cleavage parallel to the six different directions of a dodecahedron. The outlines of this figure may occasionally be detected in pieces of sphalerite within sections due to the intersections of inclined cleavage planes. In fragments sometimes almost perfect dodecahedrons may be observed.

Tendency to Break in Elongate Directions.—Some minerals exhibit a decided fibrous structure, being made up of numerous small needles visible as parallel crystals beneath the microscope. These may vary in size from small elongated blades to minute capillary fibers. Coarser minerals of this sort change from fibrous to bladed shapes.

Orientation.—The optical orientation of a mineral involves the correlation of the positions of the optical directions with crystallographic directions. In the case of biaxial minerals the problem is usually resolved into locating the position of the acute bisectrix, optic normal, and axial plane with respect to the axes *a*, *b*, and *c* of a crystal. The orientation of uniaxial minerals is less difficult, involving only the relation of the optical system to the *c*-axis. In order to make such a correlation it is necessary to take into account the usual conventions concerning the crystallographic axes of reference. In the following discussion it is assumed that the reader is familiar with the simple rules of description of crystals in the various systems and the conventions of orientation; otherwise a text on geometrical crystallography should be consulted.

Isometric System.—Optical orientation in isometric crystals is simplified since isometric crystals are isotropic, hence becoming non-directional as far as light is concerned. Thus all directions in isometric crystals are optically the same.

Tetragonal and Hexagonal Systems.—The optic axis of uniaxial minerals is always parallel to the *c*-axis of tetragonal or hexagonal minerals. The direction may be either the fast or the slow ray of the mineral, depending upon the optic sign, but other variation does not exist in tetragonal or hexagonal crystals.

Orthorhombic System.—The crystallographic axes a, b, and c of orthorhombic minerals correspond with the vibration axes X, Y, and Z but not necessarily in the order named. The crystallographic axis a, for instance, may be X, Y, or Z, and the same substitutions are possible for the crystallographic axes b and c. If two vibration axes are fixed, however, the third becomes known. Thus, if $a = Z$ and $b = Y$, it naturally follows that $c = X$. It is also evident that X and Z define the position of the axial plane; thus, if $a = Z$ and $c = X$, the axial plane includes a and c. The axial angle 2V may vary in amount in the axial plane.

Orthorhombic crystals are detected in a number of ways with the microscope. In the first place the extinction between crossed nicols is always parallel to a, b, or c. Thus if a, b, and c can be ascertained from some prominent cleavage or a section of a crystal, the nature of the extinction becomes known. Cleavage and crystal faces are particularly important in optical orientation. Orthorhombic crystals with good dispersion are also useful.

When the positions of the axes a, b, and c are once ascertained, the interference figure will furnish criteria for the proper location of X, Y, and Z with respect to the crystallographic axes. The position of the optic axes will also be ascertained at the same time. In practice each mineral presents a special problem in orientation, and it is possible only to outline the general procedure in advance.

Barite furnishes a fairly good illustration of the problem involved in the optical orientation of an orthorhombic mineral. An idealized cleavage of barite is shown in Fig. 131, together with a diagram of a thin section and three oriented sections. In thin section the barite grains exhibit cleavage in three directions, the cleavage parallel to {001} being more pronounced. The grain marked R in the thin section happens to be in a position in which the angles of cleavage measure 78°22'. The c-axis is perpendicular to this section. The planes {110} and {1$\bar{1}$0} are parallel to the c-axis. In this grain the a-axis would bisect the obtuse angle of cleavage, and the b-axis would bisect the acute angle of cleavage. Between crossed nicols the extinction will be parallel to a and b, or symmetrical with respect to the cleavage. Grain S

would have parallel extinction but might be normal to the *b*-axis and not in a position to give an interference figure of the acute bisectrix type. An interference figure is oriented with respect to the cleavage as shown in grain *Q*. A test with the quartz wedge will determine the fact that the mineral is positive; hence $Bx_a = Z$. The optic normal is Y, and $Bx_o = X$. If one refers

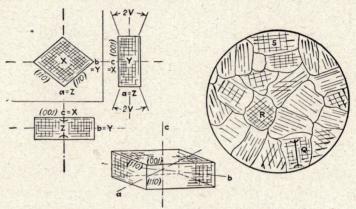

Fig. 131.—Thin-section and orientation diagrams of barite.

again to the figures illustrating the cleavage form, the following orientation is apparent:

$$a = Z$$
$$b = Y$$
$$c = X$$

The angle 2V is estimated with the microscope as approximately equal to the recorded angle 37°30′. Therefore the optic axes make an angle of 18°45′ with the *a*-axis.

The orientation in the case of orthorhombic crystals is not always so simple as in the case of barite. The principles and procedure, however, are essentially the same, and it is always fundamental to be able to fix the position of X, Y, and Z with respect to *a*, *b*, and *c*.

Monoclinic System.—In monoclinic crystals, X, Y, or Z corresponds to the *b*-axis. If Y corresponds to *b*, which is often the case, X and Z will occupy any position at 90° to each other in the plane of *a* and *c*.

Hornblende furnishes a good illustration of the problem of orientation of a monoclinic crystal. The mineral has two prominent directions of cleavage parallel to the rhombic prism {110}. The c-axis is parallel to the edge between the cleavages, and the b-axis bisects the angle between (110) and ($1\bar{1}0$) (Fig. 132). In thin sections either one or two sets of cleavage lines appear, depending upon the orientation. Grains with two cleavage directions are symmetrical in extinction and yield biaxial negative interference figures. The axial plane bisects the obtuse angle of the cleavage, and Y bisects the acute angle. The position of Z may be obtained from a section parallel to the plane of

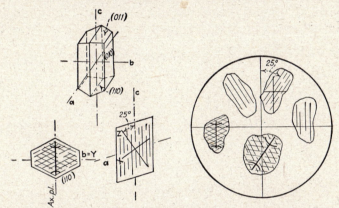

Fig. 132.—Thin-section and orientation diagrams of hornblende.

the axes a and c. Light vibrating parallel to Z is the slow ray; the angle of maximum extinction for the slow ray may be determined with the mica plate. The maximum extinction angle for the single cleavage trace is the angle between Z and the c-axis. In hornblende this angle is about 25°. When Z is determined, the angle of X is known since it lies at 90° to Z. Y is perpendicular to the plane of Z and X.

Triclinic System.—Each triclinic crystal constitutes an individual case in optical orientation. The center of the optical system must coincide with the center of the crystallographic system; otherwise there is no agreement.

Suggested References

GROUT, F. F.: "Petrography and Petrology," McGraw-Hill Book Company, Inc., New York, 1932.

Harker, A.: "Petrology for Students," 7th ed., Cambridge University Press, London, 1935.

Kraus, E. H., W. F. Hunt, and L. S. Ramsdell: "Mineralogy," 3d ed., McGraw-Hill Book Company, Inc., New York, 1936.

Johannsen, A.: "Rock-forming Minerals in Thin Sections," John Wiley & Sons, Inc., New York, 1908.

Larsen, E. S., and H. Berman: Microscopic Determination of the Non-opaque Minerals, 2d ed., *U. S. Geol. Survey Bull.* 848, 1934.

Luquer, L. M.: "Minerals in Rock Sections," 4th ed., D. Van Nostrand Company, Inc., New York, 1913.

Rogers, A. F.: "Introduction to the Study of Minerals," 3d ed., McGraw-Hill Book Company, Inc., New York, 1937.

Weinschenck, E.: "Petrographic Methods," trans. by R. W. Clark, McGraw-Hill Book Company, Inc., New York, 1912.   (Out of print.)

Winchell, A. N.: "Elements of Optical Mineralogy," 2d ed., Part II: Descriptions of Minerals, John Wiley & Sons, Inc., New York, 1927.

CHAPTER VIII

OBSERVATION OF MINERAL FRAGMENTS

Crushed Fragments.—The principles of microscopic observation with polarized light apply equally whether minerals are studied in thin sections or in crushed fragments. However, certain inherent differences in the methods of mounting, the cleavage development, or the distribution and orientation of the minerals in the mount enable one to make a few optical determinations more accurately and conveniently with fragments than with thin sections. As a result, thin-section studies should be supplemented by the observation of crushed fragments.

In the examination of crushed fragments a small amount of material suffices for a complete examination; no preliminary grinding is necessary; a minimum of equipment is used (polarizing microscope and set of index liquids); cleavage fragments break according to form and assist greatly in establishing orientation; and individual crystals may be isolated for independent examination. Nevertheless, in spite of the utility of fragment methods, it should be kept in mind that this type of examination fails to yield the splendid exhibit of mineral association and texture observable in thin sections.

The practical utilization of crushed fragments in mineral identification was first attempted by Maschke in 1872, although it was not until years later that the method was sufficiently developed to be widely used. In 1898, Schroeder van der Kolk assembled a list of fluids suitable for use as immersion liquids; in 1906, his publication of a list giving a more extensive series of liquids, together with the refractive indices of 300 minerals determined by comparison with these liquids, was a great step forward. Rogers (1906) emphasized the importance of cleavage and orientation in fragments in applying the immersion method. In recent years, E. S. Larsen, H. E. Merwin, C. S. Ross, F. E. Wright, H. Berman, and others have accumulated data that have greatly increased the application of the method.   Perhaps the greatest stimulus has been brought about by Larsen's "Microscopic Determination of the Non-opaque Minerals."

Methods of Mounting.—Permanent mounts are prepared with Canada balsam.   In the preparation of the mounts, fragments are sprinkled on a slide and covered with balsam; then the balsam is cooked by placing the slide on a hot plate, and the mount is covered with a cover glass.   This may be done in a few minutes, and the slide may then be marked for purposes of filing.

Temporary mounts are made by utilizing the various inert liquids used for determining the indices of refraction.   Screened fragments are preferable for routine work.   If screens are not available, fragments of suitable size may be obtained by placing small lumps of a mineral between two glass slides and rubbing the slides together until the glass surfaces slide smoothly over each other on the mineral powder.   On a clean glass slide, place about as many fragments as can be picked up on the tip of a small knife blade.   By tapping the edge of the slide with the knife, distribute the fragments evenly over an area approximately ¼ in. in diameter.   Then place a cover glass over the grains. With a small glass rod, apply a drop of oil at the edge of the cover glass.   The oil will be drawn in and around the grains by capillary action.   If one drop of oil is not sufficient, a second and a third may be applied until the fragments are completely immersed. Care should be taken, however, not to apply an excess of oil for this may cause the grains to float, thus hindering observations and measurements.   With sufficient care, two or three mounts may be made on a single slide.   After use, the glass slides may be cleaned with alcohol, followed by rinsing in water and drying.

Temporary mounts are by no means restricted to fine powders. Small, well-formed crystals, detrital grains, coarse fragments and oriented cleavages or sections are representative of materials that may be examined in this way.   It is even possible by skillful manipulation to turn a mineral fragment in such a way that it may be examined in a variety of positions under the microscope. A slight movement of the cover glass on a temporary mount frequently suffices to turn a fragment.   This is somewhat easier if powdered glass slightly coarser than the fragments being studied is mixed with the sample.   The glass fragments being free from cleavage lie in random positions and are easily turned, moving the fragments of the sample at the same time. The glass is easily distinguished by its isotropic character and index of refraction.

A binocular microscope of low power is extremely useful for the purpose of isolating small crystals or mineral grains (Fig. 133). While looking through the binocular small crystals may be pried loose from a specimen with a needle and these form ideal samples for examination with the polarizing microscope, in either temporary or permanent mounts.

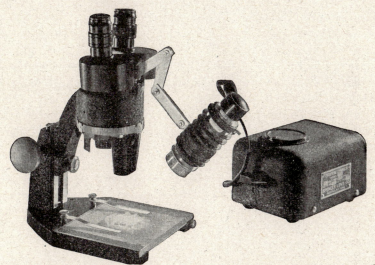

FIG. 133.—A binocular microscope.   (*Courtesy of Spencer Lens Company.*)

Immersion Method.—In immersion, mineral fragments are placed under the microscope in a liquid of known index, and the refraction of light observed at the edges of the fragments may be utilized in index comparison.

The fragments examined may vary widely in size. Ordinarily, fragments that pass through a standard 100-mesh screen and are caught on a 120-mesh screen are satisfactory. Refractive index determinations of minerals follow a routine procedure. Since the index of refraction of the liquid is known, the method of central illumination or that of oblique illumination may be used to compare the indices of refraction of the mineral fragments with the index of refraction of the liquid. As soon as the index of the mineral is determined to be greater or less than the liquid, the slide is placed in a cleaning jar, and another mount is prepared, using a liquid with a different index. By utilizing a number of liquids and making repeated comparisons the indices of refraction of a mineral may be ascertained within reasonably narrow limits. In case the mineral is isotropic, having only one index of refraction, it is customary to determine the index of the mineral to within two or three in the third place of decimals. When minerals are anisotropic, having indices of refraction that vary with direction, equal accuracy is possible, but more comparisons are necessary to determine different values. It is customary to prepare a set of liquids with indices of refraction of 1.450, 1.455, 1.460, 1.465 . . . etc., up to a limit of 1.740. A considerable number of bottles are required for such a group of liquids. These are usually kept in a dark place and are painted black on the outside to keep out the light. A less elaborate set of liquids with values 1.50, 1.51, 1.52, 1.53 . . . etc., is frequently used. Such a set requires half the number of bottles, is more quickly standardized, and with careful mixing of drops on a slide will yield results accurate to ± 0.003.

It is not always advisable to place the cover glass over dry fragments and allow the liquid to be drawn around each grain by capillary attraction, as explained above. To cite a specific instance, it may be desired to find the index of refraction of a mineral with a value between two liquids such as 1.555 and 1.560. In this case a drop of liquid 1.555 is placed upon a slide, and a drop of liquid 1.560 of the same size is placed near by but not quite touching the first drop. The mineral powder is then sprinkled between the drops, and both the drops and the powder are mixed with a needle, small wire, or small stirring rod. A small cover glass is then placed over the mixture, pressed down to remove air globules, and the mount is ready for microscopic examination. The index of the mixed liquid should be about 1.5575.

Index Determinations by Immersion.—For hexagonal or tetragonal minerals n_ϵ and n_ω are determined; for orthorhombic, monoclinic, or triclinic minerals n_α, n_β, and n_γ. In these determinations it should be remembered that each mineral grain in the field of the microscope is doubly refracting with values n_2 and n_1. Furthermore, when a crystal is in the position of extinction, the planes of vibration of n_2 and n_1 are parallel to the vibration planes of the nicols.

The determination of n_ϵ and n_ω, or n_α and n_γ usually resolves itself into a search with successive liquids until both the upper and lower limits of n_2 and n_1 for a given mineral are located. The difference between the two values equals the maximum double refraction of the mineral. It should be remembered, however, that cleavage or structure may limit the positions in

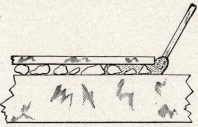

Fig. 134.—Immersion of mineral grains.

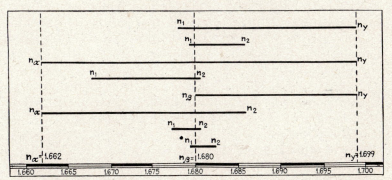

Fig. 135.—The determination of n_β in irregular non-oriented fragments.

which minerals will come to rest on the slide. In such cases the two extremes may not be obtained unless fragments are turned by moving the cover glass or interference figures are used to check the orientation.

The determination of n_β in biaxial minerals may be carried out in two ways. If interference figures on or near a bisectrix or an optic axis can be obtained for several grains in different liquids, the determination of n_β is simple. This is due to the fact that the index of the ray vibrating at right angles to the axial plane is n_β. It is also convenient to remember that any grain in a position to give an optic-axis figure yields rays with the index n_β in all directions. If such interference figures are not obtainable, and specific information concerning the position of n_β in fragments is not known, the value can still be determined by a process of elimination as explained below.

For each fragment shown in the field of the microscope, somewhere between n_2 and n_1 is the value n_β. In certain fragments either n_2 or n_1 may equal n_β, but from the consideration of biaxial crystals it can be shown that both n_2 and n_1 will never at the same time either exceed or be less than n_β in the same fragment. Thus n_β may be located by varying the liquids and observing both n_2 and n_1 in a number of fragments. In Fig. 135 assume the vertical lines to indicate the measured grains of a certain mineral. The horizontal lines cover the range in indices of refraction for different grains having different values of n_2 and n_1. It will be observed that all lines cross or meet 1.680, this being the value of n_β; also, that no lines exceed 1.699, which is the value of n_γ; and that no lines are less than 1.662, which is the value of n_α. If fragments with weak double refraction are tried in successive liquids, testing both n_2 and n_1 in each case, the value of n_β can soon be approximated within narrow limits.

It is convenient as a confirmation of the determination of optic sign to remember that when $(n_\beta - n_\alpha)$ is decidedly greater than $(n_\gamma - n_\beta)$, the mineral is optically negative. If, on the other hand, $(n_\beta - n_\alpha)$ is decidedly less than $(n_\gamma - n_\beta)$, the mineral is positive.

Form of Mineral Fragments.—Mineral fragments often exhibit distinctive forms under the microscope that are extremely useful in their identification. Such forms may be due to cleavage of the crystals of the mineral or may be due to a characteristic growth that causes peculiar fragments, or the crystals may be so small that they appear as individuals beneath the microscope.

Form of minerals in crushed fragments is probably most frequently related to the cleavage. Since cleavage is the ability of a mineral to break into smaller and smaller particles with smooth surfaces that are parallel to possible crystal faces, it follows that this characteristic will be as evident in fragments as in a hand specimen. Minerals may exhibit cleavage in one, two, three, and even four and six directions (Fig. 136). The identification of cleavage planes in mineral fragments is aided by the fact that the fragments normally orient themselves with one of the cleavage surfaces parallel to the surface of the slide. Thus, if a mineral has one direction of prominent cleavage, it will normally yield flat-lying fragments, with ragged or broken edges. Under

Fig. 136.—Cleavage and fracture fragments. (*a*) One direction—irregular plates (not illustrated). (*b*) Two directions—orthoclase, augite, and hornblende. (*c*) Three directions—kyanite, anhydrite (rectangular), calcite (rhombohedral). (*d*) Four directions—fluorite (octahedral). (*e*) Prismatic sillimanite. (*f*) Acicular—natrolite.

crossed nicols, if the mineral is anisotropic, the flat surface may exhibit a single interference color, and the narrow edges will show color bands. If the mineral has two directions of cleavage, one direction will probably be the surface on which it lies, and the other direction will show as either inclined or vertical parallel edges. If the cleavage is developed at right angles, the edge may be vertical. If, however, the cleavage develops at an inclined angle, the edge will appear beveled.

The influence of cleavage or other directional separation along smooth planes is shown in Fig. 136. The diagrams showing the cleavage fragments are idealized. In a field containing many broken fragments, a few will usually be found unmistakably representing the idealized shape. The majority, however, show only portions of the maximum cleavage development. Even minerals known to possess excellent cleavage are likely to yield a considerable number of irregular fragments when broken.

In addition to the flat-lying, inclined, or vertical planes of cleavage that modify mineral fragments, shapes due to original characteristics of crystallization may be observed. Asbestos, for example, produces threadlike fibers. Acicular or needlelike mineral structures will frequently yield fragments made up of bundles of elongated, thin, parallel crystals. Fine mica flakes may exhibit both flat-lying flakes and needlelike forms when the flakes are on edge. When the flakes are inelastic, wavy or curved plates are likely to result.

Knowledge of cleavage in fragments is necessary in using the immersion method for determining refractive indices.

Immersion Media.—Liquids for use as immersion media should be colorless, as odorless as possible, chemically stable, and miscible in all proportions with each other. They should have low dispersion, low volatility, and moderate viscosity. Liquids should be inert and not react with or dissolve the substances to be tested.

Although many different liquids have been suggested as satisfactory immersion media, it has been found that a few well-chosen liquids are preferable to a wide range of complicated compounds.[1]

Common liquids for immersion media are as shown in the table on page 143.

In the preparation of the liquids, isoamyl isovalerate and kerosene may be mixed to form liquids up to 1.466 in indices of refraction. Kerosene and halowax oil may be used for mixtures between 1.466 and 1.63. Halowax oil and methylene iodide

---

[1] Liquids suitable for index refraction determinations by the immersion method may be secured from: Geological Research Service, P. O. Box 99, Station H, New York, N. Y.; J. T. Rooney, P. O. Box 358, Buffalo, N. Y.

	Approximate Index
Liquid	of Refraction
Water	1.333
Isoamyl isovalerate[1]	1.428
Kerosene[2]	1.466
Petroleum oil	1.475
α-monobromnaphthalene[3]	1.658
Methylene iodide[4]	1.740
Solution of methylene iodide and sulfur up to	1.794

[1] Eastman Kodak Co., Rochester, N. Y.

[2] A clear, highly refined product, called *government oil*, having the value given above is sold by Leeds & Northrup Co.

[3] Monochlornaphthalene, or "halowax oil" (n = 1.63), has been found by a number of workers to be equally satisfactory and more economical.   It is sold by the Bakelite Corp.

[4] Edcan Laboratories, 12 Pine St., South Norwalk, Conn.

may be mixed to form liquids with indices of refraction between 1.63 and 1.74.   For indices below 1.43, some difficulty has been encountered in finding suitable liquids.   With more than forty minerals and many inorganic substances having indices below 1.43, liquids in this range are desirable.   Water and the alcohols are impractical because of their solvent action.   Experiments with petroleum distillates have shown that fractionation of two ligroins, gasolene, and kerosene between narrow boiling-point limits results in stable liquids with indices between 1.3548 and 1.4593 at 22° C.

It has been found also that ethyl propionate and mesitylene are miscible in all proportions to form satisfactory index media between 1.385 and 1.43.

Satisfactory liquids with indices of refraction above 1.74 are hard to obtain.   Methylene iodide containing dissolved sulfur may be used for the range from 1.74 to 1.78.   Phenyldiodoarsine (n = 1.84) mixed with methylene iodide may be used for the range from 1.78 to 1.84.   A set of high index liquids made up of phosphorus, sulfur, and methylene iodide covering the range 1.78–2.06 has been described by C. D. West.   These liquids are practically stable, inexpensive, and safe to use with the proper precautions.

Another set of liquids, composed of arsenic bromide-arsenic sulfide-methylene iodide, has been described by L. H. Borgström. These liquids have a range from 1.78 to 1.95.

Mixed melts of sulfur and selenium (2.05 to 2.72) or piperine and arsenic and antimony triiodides (1.68 to 2.10) are used for high index determinations.   The melts are prepared in advance

and arranged in a series of mixtures. The values of the indices of
the mixtures are determined by the prism method. A small
prism of the transparent melt is made and mounted on the stage
of a goniometer. Both the angle of the prism and the angle
of minimum deviation are measured. The index of refraction
is then computed as previously explained for the prism method.
The melt is molded into the form of a prism by using two cover
glasses placed at an angle of 30° to each other. The melt is
poured into the space between and allowed to cool. When cold,
the cover glasses will break away, leaving a prism with smooth
surfaces. Great care must be used not to overheat, or the indices
of the piperine mixtures will vary greatly. When the index of
the melt has been determined, the material is placed in a sealed
tube and filed for future use. When applied, the ground-up
material is melted around fragments of minerals, and the indices
are compared by the method of central illumination.

Standardization and Care of Liquids.—Indices of refraction of
standard liquids are determined according to the methods of
refractive-index determination already mentioned. Standard
refractometers are suitable for indices to 1.7; the goniometer and
prism may be used for the entire range, including melts. Liquids
should be standardized for the temperature at which they are to
be used, since changes in the index of a liquid occur with temperature variations. Generally, increase in temperature lowers
the index of a liquid. This change is approximately 0.0004 per
degree centigrade for liquids with indices below 1.658, but for
methylene iodide, it amounts to as much as 0.0007 for each
degree.

In preparing a liquid of a desired index by mixing two liquids
of known indices, the following formula is convenient:

$$V_1 n_1 + V_2 n_2 = V_x n_x$$

where V represents volume, n the index, and $V_x n_x$ the volume and
index desired. To illustrate, let the volume of $n_1 = V_1$ and the
volume of $n_2 = V_2$, $n_1 = 1.4$, $n_2 = 1.65$ and $V_x n_x = 20$ cc. of
index 1.5. Two equations may be written as follows:

$$1.4 V_1 + 1.65 V_2 = 30$$
$$V_1 + V_2 = 20$$

and, by solving,

$$V_2 = 8$$
$$V_1 = 12$$

Therefore, 8 cc. of index liquid 1.65 and 12 cc. of liquid 1.4 will make 20 cc. of a liquid with an index of 1.5. For methylene iodide mixtures, it has been found that slightly more methylene iodide may be required than the formula indicates.

For accurate work it is important that the liquids receive the best care. Black bottles (from 15 to 20 cc.) with glass applicator stoppers and glass caps are preferable for storage. A wooden box, fitted with a block recessed with round holes into which the bottles fit securely, makes a convenient and safe container for a complete set. While the liquids are in use, care should be taken to keep the bottles stoppered except when the applicator is actually being used.

Suggested References

BORGSTRÖM, L. H.: Contribution to the Development of the Immersion Method, *Bull. comm. géol. Finlande* 87, pp. 58–63, 1929.

GLASS, J. J.: Standardization of Index Liquids, *Am. Mineralog.*, vol. 19, pp. 459–465, 1934.

KAISER, E. P., and W. PARRISH: Preparation of Immersion Liquids, *Ind. Eng. Chem.*, anal. ed., vol. 11, pp. 560–562, 1939.

LARSEN, E. S., and H. BERMAN: The Microscopic Determination of the Non-opaque Minerals, *U. S. Geol. Survey Bull.* 848, 1934.

MASCHKE, O.: Ueber Abscheidung krystallisirter Kieselsäure aus wässrigen Lösungen, *Pogg. Ann.*, 5th ser., vol. 25, pp. 549–578, 1872.

MERWIN, H. E.: Media of High Refraction, etc., *Am. Jour. Sci.*, vol. 34, pp. 42–47, 1912.

————: Media of Lower Refraction, *Jour. Wash. Acad. Sci.*, vol. 3, pp. 35–40, 1913.

ROGERS, A. F.: The Determination of Minerals in Crushed Fragments by Means of the Polarizing Microscope, *School Mines Quart.*, vol. 27, pp. 340–359, 1906.

SCHROEDER VAN DER KOLK, J. L. C.: "Tabellen zur mikroskopischen Bestimmung der Mineralien nach ihrem Brechungsindex," Wiesbaden, 1906.

WEST, C. D.: Immersion Liquids of High Refractive Index, *Am. Mineralog.*, vol. 21, pp. 245–249, 1936.

WRIGHT, F. E.: The Methods of Petrographic-microscopic Research, *Carnegie Inst. Pub.* 158, 1911.

CHAPTER IX

PROCEDURE FOR THE IDENTIFICATION OF MINERALS IN THIN SECTIONS

Summary of a Scheme for Identification.—The following steps are outlined to assist in rapid identification. It is essential in identifying minerals with the microscope to follow some system that is not too involved, and at the same time effort should not be wasted on a large number of tests having no application in identification.

A mineral is seldom identified by means of one simple test. Usually a number of observations become necessary before the identification is complete. Each identification is based upon a number of characteristics determined by study and tabulation of results. The conclusion reached is always based upon a systematic analysis of the observations.

Tables are included to assist in the analysis of observations. The tables are arranged as far as possible to permit a direct interpretation in terms of possible minerals after making a few definite tests with the microscope. The tables merely offer suggestions, however, and final decision in each case should be withheld until the text has been consulted.

Opaque Minerals.—A few common opaque minerals are listed in Table I. If the mineral is transparent, follow the sequence of tables under transparent minerals as shown in the Key to Mineral Tables (page 148).

Color.—If the mineral is colored, consult Table II for colored minerals. Note also whether the mineral is pleochroic or non-pleochroic and the colors of the pleochroism if pleochroic.

Shape or Form.—If the mineral occurs in some suggestive or significant shape, consult Table III. A number of common forms or shapes are illustrated in the table.

Cleavage.—If traces of cleavage are observed, consult Table IV. Various types of cleavage and their examples among minerals are listed in this table.

Index of Refraction.—Since the thin sections are mounted in balsam, comparison of the index with balsam is possible whenever the mineral is in contact with balsam on the edge of the section or bordering a hole in the slice (balsam, $n = 1.537$). The comparative indices of adjacent minerals should also be observed.

The intensity of the boundaries and irregularities of surface for minerals in thin sections varies through wide limits. This property, known as *relief*, should be estimated when the index of refraction is compared with balsam. In most instances, by comparison with minerals of known relief the relief of an unknown mineral may be estimated as low, moderate, high, or extreme.

The identification of minerals studied in thin sections is frequently greatly facilitated and the certainty of identification increased by isolating a small portion of a puzzling mineral from the original specimen and determining the exact indices of refraction by the immersion method. Additional data concerning cleavage and orientation are also occasionally obtained in this way.

In order to be of assistance for such determinations, the indices of refraction of the minerals included in the text are recorded in Table V.

Isotropic or Anisotropic.—The group of isotropic minerals is small, and perhaps the mineral may be recognized directly if color, cleavage, and the relative index of refraction are taken into account. Isotropic minerals are listed in Table VI.

Double Refraction or Birefringence.—Anisotropic minerals vary in double refraction through wide limits. In general, however, the order of the highest observed interference color for an unknown mineral can be determined within limits of accuracy sufficient to be of considerable assistance in determining a mineral. Since properly ground thin sections are about 0.03 to 0.035 mm. in thickness, the double refraction may be estimated from the interference colors and reference to the interference color chart and Table VII.

Extinction and Elongation.—When a mineral shows cleavage or occurs in such a form that definite knowledge of orientation is possible, the angle of extinction may be utilized in identification. Elongated crystals may also be tested for either positive or negative elongation.

Optic Sign or Optical Character.—The optical classification of anisotropic minerals as uniaxial or biaxial, positive or negative, may conveniently be applied with the aid of interference figures. Since each of the four groups contains a large number of minerals, other properties as outlined in the other tables must be relied upon in addition before the identification is restricted to an individual mineral. Uniaxial positive and negative minerals are listed in Table VIII. Biaxial positive and negative minerals are listed in Tables IX and X.

Optic Angle.—In addition to the determination of optic sign, it is possible in the case of biaxial minerals to utilize the optic angle 2V or the observed angle 2E in identification.

Dispersion.—A considerable number of biaxial minerals also exhibit dispersion of the optic axes, and the dispersion formula $r > v$ or $r < v$ may be used in such cases.

Suggested Outline.—It is suggested that the systematic outline on page 172 should be followed in the identification of each mineral. It is not necessary to apply each specific test for every mineral. Often a mineral may be identified by means of a comparatively few observations. The procedure should be of value, however, in cases of doubt.

KEY TO MINERAL TABLES

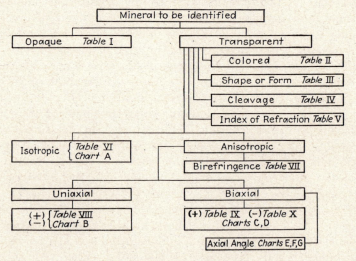

TABLE I.—OPAQUE MINERALS

Color with reflected light	Mineral	Comment
Black..............	Graphite, C	Frequently occurs in thin flakes or scattered specks with micaceous cleavage. Carbonaceous matter may occur as fine black inclusions
Brass yellow........	Pyrite, FeS_2	Euhedral crystals are common with square, triangular, or rectangular sections. Cleavage indistinct or absent. High relief due to hardness
Bronze to copper red	Pyrrhotite, $FeS(S)$.	Found in masses or bladed crystals. Basal parting ∥ to {0001}. Darker color than pyrite
Strong brass yellow..	Chalcopyrite, $CuFeS_2$	Found in masses and occasionally in euhedral crystals. Deeper yellow than pyrite and lower relief
Steel blue, red, or black.	Hematite, Fe_2O_3	Blood red on translucent edges. Occasionally shows parting. Found in euhedral crystals and masses
Violet black........	Ilmenite, $FeTiO_3$	More violet or purple than hematite. Basal parting and flake like crystals common. Often found with magnetite
Steel blue black.....	Magnetite $Fe^{II}Fe_2^{III}O_4$	Octahedral (occasionally dodecahedral) crystals are common. The mineral may have octahedral parting. Frequently a primary mineral
Iron black to brownish black.	Chromite $(Fe,Mg)(Cr,Al,Fe)_2O_4$	Usually brown on thin edges. Frequently found with serpentine
Yellow brown.......	Limonite, $H_2Fe_2O_4(H_2O)_x$	Colloform, pisolitic, porous, and massive aggregates common. Thin, transparent edges are dark between crossed nicols
White..............	Leucoxene, TiO_2, etc.	A white, fine-grained opaque alteration product of primary titanium minerals

TABLE II.—COL

			Red	Pink rose	Orange	Brown	Yellow
Isotropic non-pleochroic	Isometric		*Sphalerite* *Spinel* Cliachite Perovskite	*Sodalite* *Fluorite* Perovskite Garnet	 Perovskite	Sphalerite *Fluorite* Spinel Collophane Cliachite Perovskite Garnet	*Sodalite* Sphalerite *Fluorite* *Spinel* Collophane *Garnet*
Anisotropic pleochroic[1]	Uniaxial	(+)	Rutile			Zircon Cassiterite Rutile	Zircon *Cassiterite* *Rutile* [*Chloritoid*]
		(−)		*Tourmaline* *Corundum*		[*Biotite*] Dravite Tourmaline *Apatite*	[*Biotite*] Dravite Schorlite *Tourmaline* Jarosite
	Biaxial	(+)	Piedmontite Perovskite	*Piedmontite* Titanite *Staurolite* *Clinochlore* Perovskite	*Piedmontite* *Staurolite* Perovskite	 *Titanite* *Monazite* Aegirine-augite Chondrodite Perovskite	Piedmontite Titanite Staurolite Monazite Chloritoid *Clinochlore* Aegirine-augite Chondrodite
		(±)	Iddingsite			Iddingsite	
		(−)	*Allanite*	 Hypersthene Andalusite Dumortierite		Biotite Allanite Phlogopite Basaltic Hornblende *Aegirine* *Hypersthene* Hornblende	Biotite Epidote *Glaucophane* Allanite Phlogopite Actinolite *Glauconite*

Italics indicate lesser examples; brackets, occasional examples.

[1] Colored minerals, although normally pleochroic, may fail to vary in color with rotation

ORED MINERALS

Green	Blue	Violet	Gray	Black
	Sodalite		*Sodalite*	
Fluorite Spinel	Fluorite Spinel	Fluorite		Spinel
Perovskite *Garnet*		*Perovskite*	*Cliachite* *Perovskite*	*Garnet*
	Lazurite			
Rutile [*Chloritoid*] [*Chlorite*]	*Rutile* [*Chloritoid*] [*Chlorite*]	*Rutile*	*Zircon* *Cassiterite* *Rutile*	
[*Biotite*]				
Schorlite *Tourmaline* *Apatite*	Schorlite Corundum *Apatite*	Schorlite Corundum	Schorlite Apatite	Schorlite
Titanite		Piedmontite	Titanite	*Titanite*
Chloritoid Chlorite Aegirine-augite	Chloritoid			
Perovskite		*Perovskite*	*Perovskite*	
Crocidolite	Crocidolite	*Crocidolite*		
Chlorite Riebeckite	Chlorite Riebeckite			
Biotite Epidote	*Epidote* Glaucophane	Glaucophane	*Glaucophane*	
Lamprobolite Aegirine Hypersthene Actinolite Glauconite				*Aegirine* *Glauconite*
Hornblende	Dumortierite	Dumortierite		
	Cordierite Lazulite			

under exceptional circumstances.

TABLE III.—FORM
Minerals Found in Euhedral Crystals
cc = very common   *c* = common   *r* = rare   *rr* = very rare

Isometric	Tetragonal	Hexagonal	Orthorhombic	Monoclinic	Triclinic
Pyrite *c*	Rutile *c*	Quartz *c*	Celestite *r*	Gibbsite *r*	Microcline *r*
Fluorite *r*	Cassiterite *c*	Corundum *c*	Forsterite *c*	Monazite *r*	Plagioclase *c*
Spinel *r*	Melilite *c*	Calcite *rr*	Olivine *c*	Lazulite *rr*	
Magnetite *c*	Idocrase *c*	Dolomite *r*	Fayalite *c*	Orthoclase *c*	
Perovskite *c*	Zircon *c*	Jarosite *rr*	Monticellite *c*	Sanidine *cc*	
Leucite *cc*	Scapolite *r*	Alunite *rr*	Topaz *r*	Adularia *c*	
Sodalite *c*		Apatite *cc*	Andalusite *r*	Aegirine-augite *c*	
Haüyne *c*		Dahllite *c*	Zoisite *r*	Spodumene *c*	
Garnet *cc*		Cancrinite *r*	Staurolite *c*	Jadeite *rr*	
Analcime *r*		Schorlite *r*	Lawsonite *r*	Lamprobolite *cc*	
		Dravite *c*	Dumortierite *r*	Sphene *c*	
		Chabazite *c.*		Epidote *c*	
		Nepheline *r*		Pyroxene *c*	
				Amphibole *c*	

Form of Individual Crystals

Equant grains	Acicular	Lathlike	Columnar
Fluorite	Rutile	Ilmenite	Quartz
Quartz	Aragonite	Aragonite	Corundum
Periclase	Sillimanite	Barite	Orthoclase
Rutile	Dumortierite	Celestite	Sanidine
Cassiterite	Tourmaline	Gypsum	Microcline
Spinel	Stilbite	Aegirine	Anorthoclase
Magnetite	Natrolite	Mullite	Plagioclase
Chromite		Dumortierite	Nepheline
Anhydrite		Tourmaline	Cancrinite
Apatite		Epidote	Pyroxene
Leucite		Piedmontite	Spodumene
Sodalite		Prehnite	Wollastonite
Haüyne		Pyrophyllite	Amphibole
Melilite		Kyanite	Glaucophane
Forsterite			Beryl
Olivine			Scapolite
Fayalite			Idocrase
Chondrodite			Topaz
Garnet			Kyanite
Zircon			Zoisite
Topaz			Clinozoisite
Andalusite			Staurolite
Axinite			Micas
Allanite			Chlorites
Cordierite			Barite
Sphene			
Lawsonite			
Glauconite			
Analcime			

TABLE III.—FORM (*Continued*)
Forms of Crystal Aggregates

Granular	Fibrous	Acicular	Lathlike	Foliated
Quartz	Brucite	Aragonite	Feldspar	Graphite
Chalcedony	Gypsum	Dumortierite	Hedenbergite	Hematite
Gibbsite	Polyhalite	Tourmaline	Jadeite	Brucite
Calcite	Jadeite	Stilbite	Wollastonite	Muscovite
Dolomite	Wollastonite	Natrolite	Tremolite-	Biotite
Magnesite	Anthophyllite	Thomsonite	actinolite	Phlogopite
Siderite	Tremolite-	Scolecite	Grunerite	Lepidolite
Barite	actinolite		Glaucophane	Prochlorite
Celestite	Cummingtonite		Beryl	Clinochlore
Anhydrite	Grunerite		Scapolite	Pennine
Gypsum	Nephrite		Topaz	Chloritoid
Polyhalite	Riebeckite		Andalusite	Anthophyllite
Alunite	Sillimanite		Tourmaline	Iddingsite
Jarosite	Prehnite		Zoisite	Talc
Dahllite	Antigorite		Clinozoisite	Pyrophyllite
Olivine	Chrysotile		Epidote	Kaolinite
Epidote	Mesolite		Piedmontite	Montmorillonite
Kaolinite	Pyrophyllite		Staurolite	Dickite
Halloysite			Biotite	Hydromuscovite
Montmorillonite			Thomsonite	
Analcime			Scolecite	
			Idocrase	
			Scapolite	
			Dumortierite	
			Aragonite	

Radiated	Spherulitic	Reticulated	Graphic intergrowths	Incipient crystals
Dahllite	Chalcedony	Gibbsite	Quartz-feldspar	Cristobalite in
Cummingtonite	Cristobalite	Antigorite	Quartz-staurolite	glass
Schorlite	Calcite		Quartz-actinolite	
Prochlorite	Siderite		Nepheline-feldspar	
Pyrophyllite	Dahllite		Corundum-andalusite	
Natrolite	Orthoclase		Glass-leucite	
Chalcedony	Prehnite			
Gibbsite				
Thomsonite				
Aragonite				
Stilbite				
Dumortierite				

TABLE III.—FORM (*Continued*)
Structures

Shatter cracks	Shards	Perlitic cracks	Lithophysae	Flow (with phenocrysts)
Glass Halloysite Opal	Glass Montmorillonite	Glass	Tridymite	Lechatelierite Glass

Banded	Colloform	Oolitic	Pisolitic	Organic structures

Lechatelierite
Dahllite
Chalcedony
Opal
Calcite
Aragonite
Barite
Fluorite
	Opal
	Siderite
	Collophane
	Halloysite
		Limonite
		Calcite
		Siderite
		Collophane
		Chamosite
		Palagonite
			Cliachite
			Limonite
				Bone (a):
				  Collophane
				Cellular (b):
				  Chalcedony
				  Opal
				  Quartz
				
				Microfossil
				  replacement
				Opal
				Calcite
				Aragonite
				Dolomite
				Collophane
				Glauconite
				Chalcedony

TABLE IV.—CLEAVAGE

P = perfect $\qquad\qquad$ D = distinct

Minerals with Cleavage in One Direction

Index	Relief	Double refraction 0.01 or less	Double refraction 0.01–0.03	Double refraction 0.03 or greater
$n < balsam$	Low	Stilbite (010)P Heulandite (010)P Thomsonite (010)D Gypsum (010)P	Montmorillonite (001)D	
$n > balsam$	Low	Kaolinite (001)D Dickite (001)D Chlorite (001)P	Cordierite (010)D Gibbsite (001)D Alunite (0001) D Brucite (0001)P Polyhalite (100)D	
	Low to moderate		Cancrinite (10$\bar{1}$0)D	Prehnite (001)D Pyrophyllite (001)P Talc (001)P Muscovite (001)P Lepidolite (001)P
	Moderate to high	Topaz (001)P Zoisite (010)P Clinozoisite (010)P Staurolite (010)D Chloritoid (001)P Corundum (0001) parting	Prehnite (001)D Glauconite (001)P Sillimanite (100)P Mullite (100)D Dumortierite (100)D	Phlogopite (001)P Biotite (001)P Chondrodite (001)P Epidote (001)P Piedmontite D Jarosite (0001)D Monazite (001) (parting)

Minerals with Cleavage in Two Directions

Index	Relief	Double refraction 0.01 or less	Double refraction 0.01–0.03	Double refraction 0.03 or greater
$n < balsam$	Low	Mesolite (110)P(1$\bar{1}$0)P Scolecite (110)D Orthoclase (001)P(010)D Sanidine (001)P(010)D Anorthoclase (001)P(010)D Microcline (001)P(010)D Albite (001)P(010)D	Natrolite (110)P	
	Low	Plagioclase (001)P(010)D	Scapolite (110)D	
$n > balsam$	Moderate to high	Enstatite (110)P Andalusite (110)D Riebeckite (110)P	Tremolite-actinolite (110)P Anthophyllite (110) Glaucophane (110)P Wollastonite (100)P(001)D Hornblende (110)P Cummingtonite (110)P Jadeite (110)P Hedenbergite (110)P Augite (110)P Enstatite (110)P Hypersthene (110)P Aegirine-augite (110)P Spodumene (110)P	Diopside (110)P Grunerite (110)P Pigeonite (110)P Aegirine (110)P Lamprobolite (110)P Rutile (110)D Sphene (parting)

TABLE IV.—CLEAVAGE (*Continued*)
Minerals with Cleavage in Three or More Directions

Index	Relief	Double refraction 0.01 or less	Double refraction 0.01–0.03	Double refraction 0.03 or greater
			THREE DIRECTIONS (MISCELLANEOUS TYPES)	
$n >$ balsam	Low relief	Halite (100)P (isotropic)	Scapolite (100)P(110)D Ext. parallel or sym. Celestite (001)P(110)D	Anhydrite (001)(010)(100)P Ext. parallel
	Moderate to high relief	Axinite (001)(130)(010)D Inclined Ext. Corundum (1011) (parting)	Barite (001)(110)P Ext. parallel or sym. Lawsonite (010)P(110)D Kyanite (100)(010)P(001) parting Inclined ext.	Diaspore (010)(210)P Ext. parallel or sym. Iddingsite (100)(010)(001)D Ext. parallel
	Extreme relief	Periclase (100)P(isotropic) Perovskite (100)P(isotropic)		Rutile (100)(110)D Ext. parallel or sym.

THREE DIRECTIONS (RHOMBOHEDRAL CARBONATES)

Calcite Dolomite Magnesite Siderite	All minerals in this group exhibit the following: Variation in relief with direction moderate to high. Very high double refraction. 0.172 (calcite)–0.234 (siderite). Usually twinned. Uniaxial negative. Cleavage rhombohedral.

SPECIAL CASES

Isotropic	$n <$ balsam $n >$ balsam	Fluorite (111) octahedral Spinel (111) octahedral parting Sphalerite (110)P dodecahedral cleavage		
Anisotropic	$n >$ balsam	Relief low	Nepheline (10$\overline{1}$0)D (hexagonal)		

TABLE V.—INDICES OF REFRACTION

The indices of refraction n, n_α, n_γ, n_ϵ, and n_ω are abbreviated n, α, γ, ϵ, and ω.

Index		Mineral	Index		Mineral
1.40–1.46	n	Opal	1.531	α	Chalcedony
1.434	n	Fluorite	1.532–1.545	α	Oligoclase
1.458–1.462	n	Lechatelierite	1.532–1.552	α	Cordierite
1.469	α	Tridymite	1.535–1.570	α	Hydromuscovite
1.47–1.63	n	Palagonite	1.536–1.541	γ	Albite
1.473–1.480	α	Natrolite	1.538–1.545	α	Talc
1.473	γ	Tridymite	1.539	γ	Chalcedony
1.478–1.485	α	Chabazite	1.539–1.570	γ	Cordierite
1.48–1.61	n	Volcanic glass	1.540–1.571	ϵ	Scapolite
1.480–1.490	γ	Chabazite	1.541–1.552	γ	Oligoclase
1.483–1.487	n	Sodalite	1.541–1.579	α	Biotite
1.484	α	Cristobalite	1.544	n	Halite
1.485–1.493	γ	Natrolite	1.5442	ω	Quartz
1.486	ϵ	Calcite	1.545–1.555	α	Andesine
1.487	n	Analcime	1.548	α	Polyhalite
1.487	γ	Cristobalite	1.549–1.561	n	Halloysite
1.492	α	Montmorillonite	1.550–1.607	ω	Scapolite
1.493–1.546	α	Chrysotile	1.551–1.562	α	Phlogopite
1.494–1.500	α	Stilbite	1.552	α	Pyrophyllite
1.496–1.499	α	Heulandite	1.552–1.562	γ	Andesine
1.496–1.500	ϵ	Cancrinite	1.5533	ϵ	Quartz
1.496–1.510	n	Haüyne	1.554–1.567	α	Gibbsite
1.500–1.508	γ	Stilbite	1.555–1.563	α	Labradorite
1.500–1.526	ϵ	Dolomite	1.555–1.564	α	Antigorite
1.50–1.57	n	Serpophite	1.556–1.570	α	Muscovite
1.501–1.505	γ	Heulandite	1.560	α	Lepidolite
1.505	α	Mesolite	1.560	α	Dickite
1.506	γ	Mesolite	1.561	α	Kaolinite
1.507–1.524	ω	Cancrinite	1.562–1.571	γ	Labradorite
1.508	α	Leucite	1.562–1.573	γ	Antigorite
1.509	γ	Leucite	1.563–1.571	α	Bytownite
1.509–1.527	ϵ	Magnesite	1.564–1.590	ϵ	Beryl
1.512	α	Scolecite	1.565–1.605	γ	Hydromuscovite
1.512–1.530	α	Thomsonite	1.566	ω	Brucite
1.513	γ	Montmorillonite	1.566	γ	Dickite
1.517–1.520	α	Sanidine	1.566	γ	Kaolinite
1.517–1.557	γ	Chrysotile	1.567	γ	Polyhalite
1.518	α	Orthoclase	1.568–1.598	ω	Beryl
1.518–1.522	α	Microcline	1.570	α	Anhydrite
1.518–1.542	γ	Thomsonite	1.57–1.61	n	Cliachite
1.519	γ	Scolecite	1.57–1.62	n	Collophane
1.520	α	Gypsum	1.571–1.575	α	Anorthite
1.522–1.536	α	Anorthoclase	1.571–1.582	γ	Bytownite
1.524–1.526	γ	Sanidine	1.571–1.588	α	Clinochlore
1.525–1.530	γ	Microcline	1.572	ω	Alunite
1.525–1.532	α	Albite	1.574–1.638	γ	Biotite
1.526	γ	Orthoclase	1.575–1.582	α	Pennine
1.527–1.543	ϵ	Nepheline	1.575–1.590	γ	Talc
1.527–1.541	γ	Anorthoclase	1.576–1.583	γ	Pennine
1.529	γ	Gypsum	1.576–1.589	γ	Gibbsite
1.530	α	Aragonite	1.576–1.597	γ	Clinochlore
1.530–1.547	ω	Nepheline	1.582–1.588	γ	Anorthite

TABLE V.—INDICES OF REFRACTION (*Continued*)

Index		Mineral	Index		Mineral
1.585	ε	Brucite	1.645–1.665	γ	Prehnite
1.588–1.658	α	Prochlorite	1.648	γ	Barite
1.590–1.612	α	Glauconite	1.650–1.665	α	Enstatite
1.592	ε	Alunite	1.650–1.698	α	Diopside
1.592–1.643	α	Chondrodite	1.651–1.668	α	Spodumene
1.593–1.611	γ	Muscovite	1.651–1.681	α	Olivine
1.596–1.633	ε	Siderite	1.652–1.698	ω	Schorlite
1.598–1.606	γ	Phlogopite	1.654	γ	Mullite
1.598–1.652	α	Anthophyllite	1.655–1.666	α	Jadeite
1.599–1.667	γ	Prochlorite	1.655–1.669	γ	Monticellite
1.600	γ	Pyrophyllite	1.657–1.661	α	Sillimanite
1.600–1.628	α	Nephrite	1.657–1.663	α	Grunerite
1.600–1.628	α	Tremolite-actinolite	1.658	ω	Calcite
1.603–1.604	α	Lazulite	1.658–1.674	γ	Enstatite
1.605	γ	Lepidolite	1.659–1.678	α	Dumortierite
1.607–1.629	α	Topaz	1.66–1.80	γ	Allanite
1.610–1.644	γ	Glauconite	1.664–1.686	γ	Cummingtonite
1.613–1.628	ε	Dravite	1.665	α	Lawsonite
1.614	γ	Anhydrite	1.667–1.688	γ	Jadeite
1.614–1.675	α	Hornblende	1.670–1.680	γ	Forsterite
1.615–1.629	ε	Elbaite	1.670–1.692	α	Lamprobolite
1.615–1.635	α	Prehnite	1.673–1.715	α	Hypersthene
1.617–1.638	γ	Topaz	1.674–1.730	α	Iddingsite
1.619–1.626	ε	Dahllite	1.677–1.681	γ	Spodumene
1.620	α	Wollastonite	1.677–1.684	γ	Sillimanite
1.621–1.655	α	Glaucophane	1.678–1.684	α	Axinite
1.621–1.670	γ	Chondrodite	1.680–1.716	ω	Dolomite
1.622	α	Celestite	1.680–1.718	α	Pigeonite
1.623–1.635	ω	Dahllite	1.680–1.745	α	Aegirine-augite
1.623–1.676	γ	Anthophyllite	1.681–1.727	γ	Diopside
1.625–1.655	γ	Nephrite	1.683–1.731	γ	Hypersthene
1.625–1.655	γ	Tremolite-actinolite	1.684	γ	Lawsonite
1.626–1.629	ε	Melilite	1.686	γ	Aragonite
1.628–1.658	ε	Schorlite	1.686–1.692	γ	Dumortierite
1.629–1.640	α	Andalusite	1.688–1.696	γ	Axinite
1.630–1.651	ε	Apatite	1.688–1.712	α	Augite
1.631	γ	Celestite	1.689–1.718	γ	Olivine
1.632–1.634	ω	Melilite	1.693	α	Riebeckite
1.632–1.655	ω	Dravite	1.693–1.760	γ	Lamprobolite
1.633–1.655	ω	Apatite	1.696–1.700	α	Zoisite
1.633–1.701	γ	Hornblende	1.697	γ	Riebeckite
1.634	γ	Wollastonite	1.699–1.717	γ	Grunerite
1.635	n	Chamosite	1.700–1.726	ω	Magnesite
1.635–1.640	α	Forsterite	1.701–1.726	ε	Idocrase
1.635–1.655	ω	Elbaite	1.702	α	Diaspore
1.636	α	Barite	1.702–1.718	γ	Zoisite
1.639–1.642	γ	Lazulite	1.705–1.732	ω	Idocrase
1.639–1.647	γ	Andalusite	1.709–1.782	γ	Aegirine-augite
1.639–1.657	α	Cummingtonite	1.710–1.723	α	Clinozoisite
1.639–1.668	γ	Glaucophane	1.712	α	Kyanite
1.64–1.77	α	Allanite	1.713–1.737	γ	Augite
1.641–1.651	α	Monticellite	1.715	ε	Jarosite
1.642	α	Mullite	1.715–1.724	α	Chloritoid

TABLE V.—INDICES OF REFRACTION (*Concluded*)

Index		Mineral	Index		Mineral
1.718–1.768	γ	Iddingsite	1.792–1.820	n	Spessartite
1.719–1.734	γ	Clinozoisite	1.805–1.835	α	Fayalite
1.719–1.744	γ	Pigeonite	1.806–1.832	γ	Piedmontite
1.72–1.78	n	Spinel	1.820	ω	Jarosite
1.720–1.734	α	Epidote	1.830–1.875	ω	Siderite
1.728	γ	Kyanite	1.837–1.849	γ	Monazite
1.731–1.737	γ	Chloritoid	1.838–1.870	n	Uvarovite
1.732–1.739	α	Hedenbergite	1.847–1.886	γ	Fayalite
1.734–1.779	γ	Epidote	1.857–1.887	n	Andradite
1.736–1.747	α	Staurolite	1.887–1.913	α	Sphene
1.736–1.763	n	Grossularite	1.925–1.931	ω	Zircon
1.738–1.760	n	Periclase	1.979–2.054	γ	Sphene
1.741–1.760	n	Pyrope	1.985–1.993	ε	Zircon
1.745–1.758	α	Piedmontite	1.996	ω	Cassiterite
1.745–1.777	α	Aegirine	2.00–2.10	n	Limonite
1.746–1.762	γ	Staurolite	2.07–2.16	n	Chromite
1.750	γ	Diaspore	2.093	ε	Cassiterite
1.751–1.757	γ	Hedenbergite	2.34–2.38	n	Perovskite
1.759–1.763	ε	Corundum	2.37–2.47	n	Sphalerite
1.767–1.772	ω	Corundum	2.603–2.616	ω	Rutile
1.778–1.815	n	Almandite	2.889–2.903	ε	Rutile
1.782–1.836	γ	Aegirine	2.94	ε	Hematite
1.786–1.800	α	Monazite	3.22	ω	Hematite

TABLE VI.—ISOTROPIC MINERALS

Relief	Mineral	Index
Moderate relief	Opal	1.40 −1.46
	Fluorite	1.434
	Lechatelierite	1.458–1.462
	Sodalite	1.483–1.487
	Analcime	1.487
	Haüyne	1.496–1.510
	——Balsam = 1.537——	
Low relief	Halite	1.544
	Halloysite	1.549–1.561
	Serpophite	1.50 −1.57
	Cliachite	1.57 −1.61
	Collophane	1.57 −1.62
Moderate to strong relief	Periclase	1.738–1.760
	Grossularite	1.736–1.763
	Pyrope	1.741–1.760
	Almandite GARNET	1.778–1.815
	Spessartite GROUP	1.792–1.820
	Uvarovite	1.838–1.870
	Andradite	1.857–1.887
Very high relief	Limonite	2.00 −2.10
	Spinel	1.72 −1.78
	Chromite	2.07 −2.16
	Perovskite	2.34 −2.38
	Sphalerite	2.37 −2.47
	Volcanic glass (mineraloid)	1.48 −1.61
	Palagonite (mineraloid)	1.47 −1.63

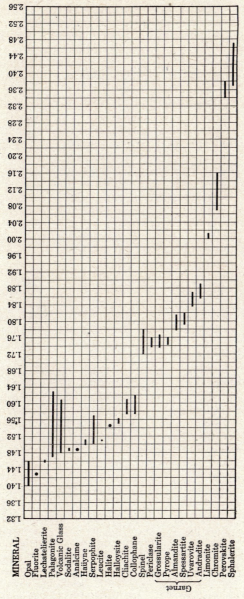

CHART A.—Range of Refractive Indices—Isotropic Minerals.

TABLE VII.—BIREFRINGENCE

Birefringence	Mineral	Birefringence	Mineral
0.00 –0.002	Analcime (possibly)	0.008–0.009	Enstatite
0.00 –0.002	Perovskite	0.008–0.011	Bytownite
0.00 –0.003	Serpophite	0.009	Celestite
0.00 –0.004	Haüyne (occas.)	0.009	Gypsum
0.001	Leucite	0.009	Quartz
0.001	Mesolite	0.009–0.010	Topaz
0.001	Halloysite	0.009–0.011	Albite
0.001–0.004	Pennine	0.010–0.012	Axinite
0.001–0.011	Prochlorite	0.010–0.015	Staurolite
0.002–0.010	Chabazite	0.010–0.016	Hypersthene
0.003	Cristobalite	0.01–0.03	Allanite
0.003–0.004	Apatite	0.010–0.036	Scapolite
0.003–0.004	Nepheline	0.011–0.013	Anorthite
0.004	Tridymite	0.011–0.014	Chrysotile
0.004	Riebeckite	0.011–0.020	Dumortierite
0.004–0.006	Idocrase	0.012	Barite
0.004–0.008	Beryl	0.012	Mullite
0.004–0.009	Dahllite	0.012–0.013	Natrolite
0.004–0.011	Clinochlore	0.012–0.023	Jadeite
0.005	Collophane	0.013–0.016	Chloritoid
0.005	Kaolinite	0.013–0.018	Glaucophane
0.005–0.006	Melilite	0.013–0.027	Spodumene
0.005–0.007	Anorthoclase	0.014	Wollastonite
0.005–0.011	Clinozoisite	0.014–0.018	Monticellite
0.006	Dickite	0.014–0.045	Epidote
0.006–0.008	Stilbite	0.015–0.023	Elbaite
0.006–0.012	Thomsonite	0.016	Kyanite
0.006–0.018	Zoisite	0.016–0.025	Anthophyllite
0.007	Sanidine	0.018–0.019	Hedenbergite
0.007	Heulandite	0.019	Brucite
0.007	Andesine	0.019	Lawsonite
0.007	Scolecite	0.019	Polyhalite
0.007	Microcline	0.019–0.025	Dravite
0.007–0.008	Chamosite	0.019–0.026	Hornblende
0.007–0.008	Labradorite	0.020	Alunite
0.007–0.009	Antigorite	0.020–0.023	Sillimanite
0.007–0.009	Oligoclase	0.020–0.032	Glauconite
0.007–0.011	Andalusite	0.020–0.033	Prehnite
0.007–0.011	Cordierite	0.021	Montmorillonite
0.007–0.028	Cancrinite	0.021–0.025	Augite
0.008	Chalcedony	0.021–0.033	Pigeonite
0.008	Orthoclase	0.022	Gibbsite
0.008–0.009	Corundum	0.022–0.027	Nephrite

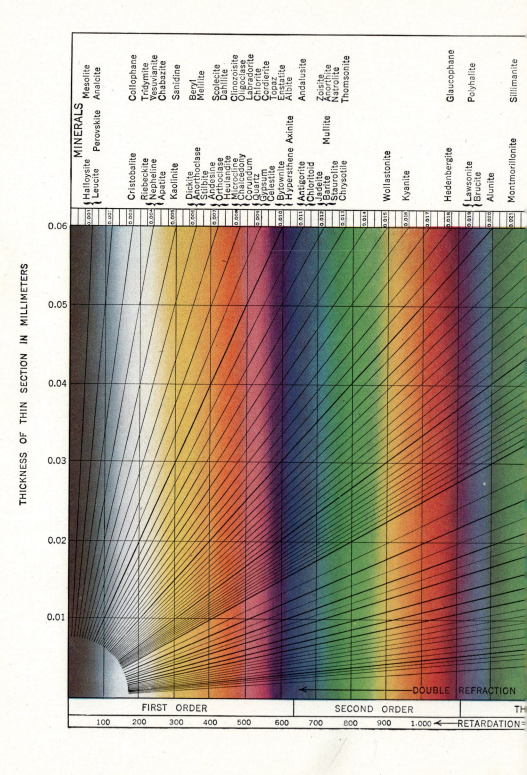

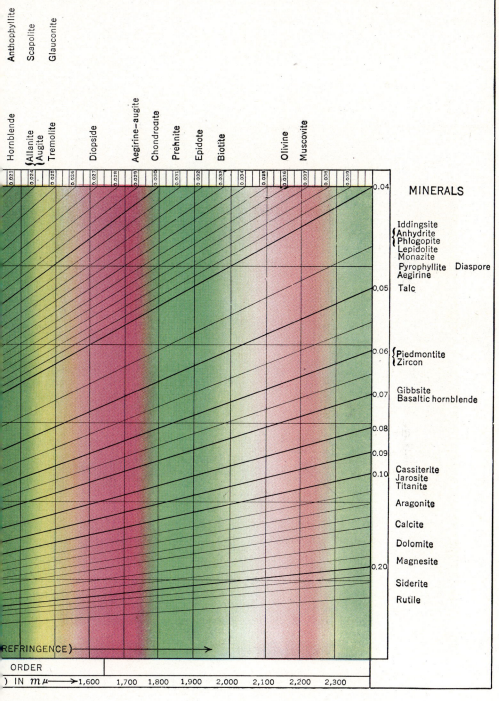

Anthophyllite
Scapolite
Glauconite

Hornblende
{Allanite
{Augite
Tremolite
Diopside
Aegirine–augite
Chondroate
Prehnite
Epidote
Biotite
Olivine
Muscovite

0.023 0.024 0.025 0.026 0.027 0.028 0.029 0.030 0.031 0.032 0.033 0.034 0.035 0.036 0.037 0.038 0.039 0.04

MINERALS

Iddingsite
{Anhydrite
{Phlogopite
Lepidolite
Monazite
Pyrophyllite   Diaspore
Aegirine
0.05   Talc

0.06 {Piedmontite
{Zircon

0.07   Gibbsite
Basaltic hornblende

0.08

0.09

0.10   Cassiterite
Jarosite
Titanite

Aragonite

Calcite

Dolomite

Magnesite

0.20   Siderite

Rutile

REFRINGENCE)

ORDER

) IN mμ ——→ 1,600   1,700   1,800   1,900   2,000   2,100   2,200   2,300

TABLE VII.—BIREFRINGENCE (*Concluded*)

Birefringence	Mineral	Birefringence	Mineral
0.022–0.027	Tremolite-actinolite	0.044	Anhydrite
0.022–0.040	Schorlite	0.044–0.047	Phlogopite
0.025–0.029	Cummingtonite	0.045	Lepidolite
0.026–0.072	Lamprobolite	0.048	Diaspore
0.027–0.035	Chondrodite	0.048	Pyrophyllite
0.029–0.031	Diopside	0.049–0.051	Monazite
0.029–0.037	Aegirine-augite	0.060–0.062	Zircon
0.030–0.035	Hydromuscovite	0.061–0.082	Piedmontite
0.030–0.050	Talc	0.092–0.141	Sphene
0.033–0.059	Biotite	0.097	Cassiterite
0.035–0.040	Forsterite	0.105	Jarosite
0.036–0.038	Lazulite	0.156	Aragonite
0.037–0.041	Olivine	0.172	Calcite
0.037–0.041	Muscovite	0.180–0.190	Dolomite
0.037–0.059	Aegirine	0.191–0.199	Magnesite
0.038–0.044	Iddingsite	0.234–0.242	Siderite
0.042–0.051	Fayalite	0.286–0.287	Rutile
0.042–0.054	Grunerite		

TABLE VIII.—UNIAXIAL MINERALS

Mineral	n_ϵ	n_ω	Sign	Birefringence
Calcite	1.486	1.658	—	0.172
Cancrinite	1.496–1.500	1.507–1.524	—	0.007–0.028
Dolomite	1.500–1.526	1.680–1.716	—	0.180–0.190
Magnesite	1.509–1.527	1.700–1.726	—	0.191–0.199
Nepheline	1.527–1.543	1.530–1.547	—	0.003–0.004
Scapolite	1.540–1.571	1.550–1.607	—	0.010–0.036
Beryl	1.564–1.590	1.568–1.598	—	0.004–0.008
Siderite	1.596–1.633	1.830–1.875	—	0.234–0.242
Dravite	1.613–1.628	1.632–1.655	—	0.019–0.025
Elbaite	1.615–1.629	1.635–1.655	—	0.015–0.023
Dahllite	1.619–1.626	1.623–1.635	—	0.004–0.009
Melilite	1.626–1.629	1.632–1.634	—	0.005–0.006
Schorlite	1.628–1.658	1.652–1.698	—	0.022–0.040
Apatite	1.630–1.651	1.633–1.655	—	0.003–0.004
Idocrase	1.701–1.726	1.705–1.732	—	0.004–0.006
Jarosite	1.715	1.820	—	0.105
Corundum	1.759–1.763	1.767–1.772	—	0.008–0.009
Hematite	2.94	3.22	—	

Mineral	n_ω	n_ϵ	Sign	Birefringence
Quartz	1.5442	1.5533	+	0.009
Brucite	1.566	1.585	+	0.019
Alunite	1.572	1.592	+	0.020
Zircon	1.925–1.931	1.985–1.993	+	0.060–0.062
Cassiterite	1.996	2.093	+	0.097
Rutile	2.603–2.616	2.889–2.903	+	0.286–0.287

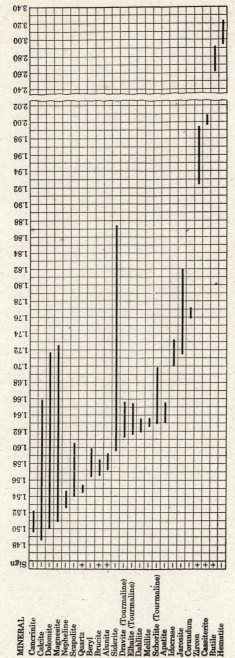

Chart B.—Range of Refractive Indices—Anisotropic Uniaxial Minerals.

TABLE IX.—BIAXIAL POSITIVE MINERALS

Mineral	n_α	n_β	n_λ	Birefringence
Tridymite	1.469	1.469	1.473	0.004
Natrolite	1.473–1.480	1.476–1.482	1.485–1.493	0.012–0.013
Chabazite	1.478–1.485	....	1.480–1.490	0.002–0.010
Chrysotile	1.493–1.546	1.504–1.550	1.517–1.557	0.011–0.014
Heulandite	1.496–1.499	1.497–1.501	1.501–1.505	0.007
Mesolite	1.505	1.505	1.506	0.001
Thomsonite	1.512–1.530	1.513–1.532	1.518–1.542	0.006–0.012
Gypsum	1.520	1.522	1.529	0.009
Albite	1.525–1.532	1.529–1.536	1.536–1.541	0.009–0.011
Oligoclase	1.532–1.545	1.536–1.548	1.541–1.552	0.007–0.009
Cordierite	1.532–1.552	1.536–1.562	1.539–1.570	0.007–0.011
Andesine	1.545–1.555	1.548–1.558	1.552–1.562	0.007
Gibbsite	1.554–1.567	1.554–1.567	1.576–1.589	0.022
Labradorite	1.555–1.563	1.558–1.567	1.562–1.571	0.007–0.008
Dickite	1.560	1.562	1.566	0.006
Anhydrite	1.570	1.576	1.614	0.044
Clinochlore	1.571–1.588	1.571–1.588	1.576–1.597	0.004–0.011
Pennine	1.575–1.582	1.576–1.582	1.576–1.583	0.001–0.004
Prochlorite	1.588–1.658	1.589–1.667	1.599–1.667	0.001–0.011
Chondrodite	1.592–1.643	1.602–1.655	1.621–1.670	0.027–0.035
Anthophyllite	1.598–1.652	1.615–1.662	1.623–1.676	0.016–0.025
Topaz	1.607–1.629	1.610–1.631	1.617–1.638	0.009–0.010
Prehnite	1.615–1.635	1.624–1.642	1.645–1.665	0.020–0.033
Celestite	1.622	1.624	1.631	0.009
Forsterite	1.635–1.640	1.651–1.660	1.670–1.680	0.035–0.040
Barite	1.636	1.637	1.648	0.012
Cummingtonite	1.639–1.657	1.645–1.669	1.664–1.686	0.025–0.029
Mullite	1.642	1.644	1.654	0.012
Enstatite	1.650–1.665	1.653–1.670	1.658–1.674	0.008–0.009
Diopside	1.650–1.698	1.657–1.706	1.681–1.727	0.029–0.031
Spodumene	1.651–1.668	1.665–1.675	1.677–1.681	0.013–0.027
Olivine	1.651–1.681	1.670–1.706	1.689–1.718	0.037–0.041
Jadeite	1.655–1.666	1.659–1.674	1.667–1.688	0.012–0.023
Sillimanite	1.657–1.661	1.658–1.670	1.677–1.684	0.020–0.023
Lawsonite	1.665	1.674	1.684	0.019
Iddingsite	1.674–1.730	1.715–1.763	1.718–1.768	0.038–0.044
Aegirine-augite	1.680–1.745	1.687–1.770	1.709–1.782	0.029–0.037
Pigeonite	1.680–1.718	1.698–1.725	1.719–1.744	0.021–0.033
Augite	1.688–1.712	1.701–1.717	1.713–1.737	0.021–0.025
Zoisite	1.696–1.700	1.696–1.703	1.702–1.718	0.006–0.018
Diaspore	1.702	1.722	1.750	0.048
Clinozoisite	1.710–1.723	1.715–1.729	1.719–1.734	0.005–0.011
Chloritoid	1.715–1.724	1.719–1.726	1.731–1.737	0.013–0.016
Hedenbergite	1.732–1.739	1.737–1.745	1.751–1.757	0.018–0.019
Staurolite	1.736–1.747	1.741–1.754	1.746–1.762	0.010–0.015
Piedmontite	1.745–1.758	1.764–1.789	1.806–1.832	0.061–0.082
Monazite	1.786–1.800	1.788–1.801	1.837–1.849	0.049–0.051
Sphene	1.887–1.913	1.894–1.921	1.979–2.054	0.092–0.141

TABLE X.—BIAXIAL NEGATIVE MINERALS

Mineral	n_α	n_β	n_λ	Birefringence
Montmorillonite	1.492	1.513	1.513	0.021
Stilbite	1.494–1.500	1.498–1.504	1.500–1.508	0.006–0.008
Scolecite	1.512	1.519	1.519	0.007
Sanidine	1.517–1.520	1.523–1.525	1.524–1.526	0.007
Orthoclase	1.518	1.524	1.526	0.008
Microcline	1.518–1.522	1.522–1.526	1.525–1.530	0.007
Anorthoclase	1.522–1.536	1.526–1.539	1.527–1.541	0.005–0.007
Aragonite	1.530	1.682	1.686	0.156
Oligoclase	1.532–1.545	1.536–1.548	1.541–1.552	0.007–0.009
Cordierite	1.532–1.552	1.536–1.562	1.539–1.570	0.007–0.011
Hydromuscovite	1.535–1.570	....	1.565–1.605	0.030–0.035
Talc	1.538–1.545	1.575–1.590	1.575–1.590	0.030–0.050
Biotite	1.541–1.579	1.574–1.638	1.574–1.638	0.033–0.059
Andesine	1.545–1.555	1.548–1.558	1.552–1.562	0.007
Polyhalite	1.548	1.562	1.567	0.019
Phlogopite	1.551–1.562	1.598–1.606	1.598–1.606	0.044–0.047
Pyrophyllite	1.552	1.588	1.600	0.048
Antigorite	1.555–1.564	1.562–1.573	1.562–1.573	0.007–0.009
Muscovite	1.556–1.570	1.587–1.607	1.593–1.611	0.037–0.041
Lepidolite	1.560	1.598	1.605	0.045
Kaolinite	1.561	1.565	1.566	0.005
Bytownite	1.563–1.571	1.567–1.577	1.571–1.582	0.008–0.011
Anorthite	1.571–1.575	1.577–1.583	1.582–1.588	0.011–0.013
Pennine	1.575–1.582	1.576–1.582	1.576–1.583	0.001–0.004
Glauconite	1.590–1.612	1.609–1.643	1.610–1.644	0.020–0.032
Tremolite-actinolite	1.600–1.628	1.613–1.644	1.625–1.655	0.022 0.027
Nephrite	1.600–1.628	1.613–1.644	1.625–1.655	0.022–0.027
Lazulite	1.603–1.604	1.632–1.633	1.639–1.642	0.036–0.038
Hornblende	1.614–1.675	1.618–1.691	1.633–1.701	0.019–0.026
Wollastonite	1.620	1.632	1.634	0.014
Chamosite	....	1.635	....	0.007–0.008
Glaucophane	1.621–1.655	1.638–1.664	1.639–1.668	0.013–0.018
Andalusite	1.629–1.640	1.633–1.644	1.639–1.647	0.007–0.011
Allanite	1.640–1.770	1.650–1.770	1.660–1.800	0.010–0.030
Monticellite	1.641–1.651	1.646–1.662	1.655–1.669	0.014–0.018
Olivine	1.651–1.681	1.670–1.706	1.689–1.718	0.037–0.041
Grunerite	1.657–1.663	1.684–1.697	1.699–1.717	0.042–0.054
Dumortierite	1.659–1.678	1.684–1.691	1.686–1.692	0.011–0.020
Lamprobolite	1.670–1.692	1.683–1.730	1.693–1.760	0.026–0.072
Hypersthene	1.673–1.715	1.678–1.728	1.683–1.731	0.010–0.016
Iddingsite	1.674–1.730	1.715–1.763	1.718–1.768	0.038–0.044
Axinite	1.678–1.684	1.685–1.692	1.688–1.696	0.010–0.012
Riebeckite	1.693	1.695	1.697	0.004
Kyanite	1.712	1.720	1.728	0.016
Epidote	1.720–1.734	1.724–1.763	1.734–1.779	0.014–0.045
Aegirine	1.745–1.777	1.770–1.823	1.782–1.836	0.037–0.059
Fayalite	1.805–1.835	1.838–1.877	1.847–1.886	0.042–0.051

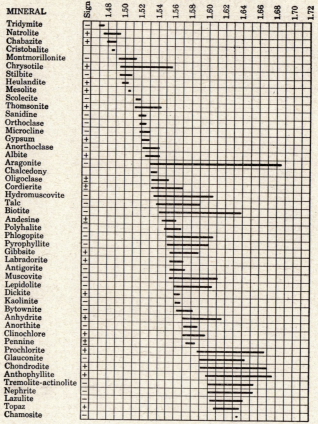

CHART C.—Range of Refractive Indices—Anisotropic Biaxial Minerals.

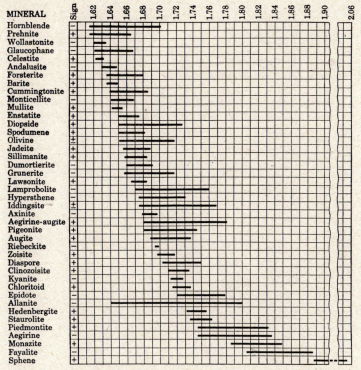

CHART D.—Range of Refractive Indices—Anisotropic Biaxial Minerals.

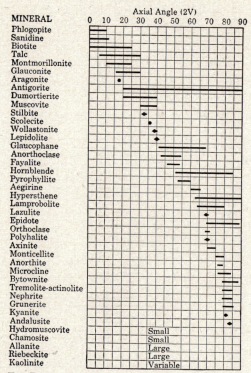

CHART E.—Range of Axial Angles—Biaxial Negative Minerals.

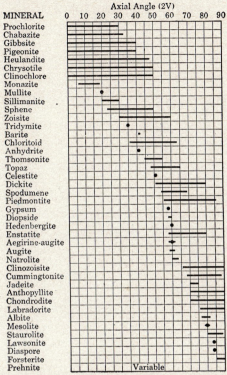

CHART F.—Range of Axial Angles—Biaxial Positive Minerals.

CHART G.—Range of Axial Angles—Biaxial Positive/Negative Minerals.

OUTLINE FOR IDENTIFICATION
Description of the mineral to be identified
....
....
Associated minerals..
   Color (if opaque)..
   Transparent...
      Properties if transparent:
         Color...
         Pleochroism...
         Shape or form...
         ....
         Cleavage..
         ....
         Indices of refraction.....................................
         ....
         Isotropic.....................or Anisotropic...............
         If anisotropic:
            Birefringence or double refraction.........................
            Twinning (if present).....................................
            Elongation (if any).......................................
         Optical classification:
            Uniaxial
               Positive..................Negative.......................
            Biaxial
               Positive..................Negative.......................
               2V.......................(or 2E).......................
               Dispersion..
         Optical orientation.......................................
         ....
Conclusion...

PART II

DESCRIPTIONS OF INDIVIDUAL MINERALS AND MINERALOIDS

INTRODUCTION TO PART II

Of the approximately 1500 known minerals, comparatively few are important rock-forming constituents. In the following pages descriptions of individual minerals and mineral groups are given. The list of minerals and mineraloids included in the descriptions and presented on page 177 includes practically all the common and important minerals found in igneous, sedimentary, and metamorphic rocks, together with the most important vein minerals. Other minerals may be important locally. Minerals other than those listed may occasionally be encountered; then the larger reference books should be consulted. A selected list of these may be found on the following page.

For convenience of reference the microscopic and optical characteristics of the minerals are given under appropriate headings. *Color* refers to the color of the mineral in thin sections. The term *neutral* is used for very pale colors of indeterminate hue. Pleochroism and absorption are included along with color, for ordinarily the lower nicol is in place during the entire examination of the slide. *Relief* refers to ordinary thin sections mounted in Canada balsam or to fragments in clove oil. The interference colors listed are usually the maximum colors for thin sections of about 0.03 mm. thickness. If sections are thicker or thinner than this value, allowance must be made. The thickness may usually be determined fairly accurately if a known mineral such as quartz or plagioclase is present in the slide. Clockwise extinction angles are considered positive; and counterclockwise ones, negative. *Orientation* refers especially to the position of the faster and slower rays in characteristic sections. The complete optical orientation is given in the tabulation just below the name of the mineral. The size of the axial angle for biaxial crystals is indicated in a general way in the text. Exact values of this angle are given in the tabulation just referred to. Approximate measurements of the axial angle can be made in favorable cases by the use of a micrometer ocular that has been calibrated by means of several biaxial crystals of known axial angles. Under the heading *Distinguishing Features* resemblances to, and differences from, similar minerals are pointed out as an aid in the determination. *Related Minerals* are the rarer minerals similar in appearance and properties to the one under discussion.

Orientation diagrams for nearly all the biaxial minerals are given, and for most of the monoclinic minerals plans combined with side elevations are inserted in order to facilitate an understanding of their optical properties.

It is to be hoped that the photomicrographs of typical thin sections will aid in the identification of minerals.

References for Part II

CHUDOBA, K.: "Mikroskopische Charakteristik der gesteinsbildenden Mineralien," Herder u. Co., Freiburg im B., 1932.

————: "Die Feldspäte und ihre praktische Bestimmung," E. Schweizerbart'sche Verlagsbuchhandlung, Stuttgart, 1932.

HATCH, F. H. and A. K. WELLS: "The Petrology of the Igneous Rocks," 9th ed., George Allen & Unwin, Ltd., London, 1937.

IDDINGS, J. P.: "Rock Minerals," 2d ed., John Wiley & Sons, Inc., New York, 1911.

JOHANNSEN, A.: "Essentials for the Microscopical Determination of Rock-forming Minerals and Rocks in Thin Sections," 2d ed., University of Chicago Press, Chicago, 1928.

LARSEN, E. S., and H. BERMAN: The Microscopic Determination of the Nonopaque Minerals, *U. S. Geol. Survey Bull.* 848, 2d ed., Washington, D. C., 1934.

MILNER, H. B.: "Sedimentary Petrography," 3d ed., Thos. Murby & Co., London, 1940.

ROSENBUSCH, H.: "Mikroskopische Physiographie der petrographisch wichtigen Mineralien," rev. by O. Mügge, Band I, 2te Hälfte, 5th ed., E. Schweizerbart'sche Verlagsbuchhandlung, Stuttgart, 1927.

TICKELL, F. G.: "The Examination of Fragmental Rocks," 2d ed., Stanford University Press, Stanford University, Calif., 1940.

WEINSCHENK, E.: "Petrographic Methods," trans. by R. W. Clark, McGraw-Hill Book Company, Inc., New York, 1912. (Out of print.)

WINCHELL, A. N.: "Elements of Optical Mineralogy," 3d ed., Part II, John Wiley & Sons, Inc., New York, 1933.

CLASSIFICATION OF THE MINERALS AND MINERALOIDS DESCRIBED

ELEMENTS
  Graphite
SULFIDS AND SULFOSALTS
  Sphalerite
  Pyrite
  Pyrrhotite
  Chalcopyrite
HALIDES
  Halite
  Fluorite
OXIDS
  SILICA MINERALS
    α-Quartz
    β-Quartz
    Chalcedony
    Opal
    Tridymite
    Cristobalite
    Lechatelierite
  Periclase
  { Corundum
  { Hematite
   Ilmenite
  { Rutile
  { Cassiterite
ALUMINATES, etc.
  Spinel
  Magnetite
  Chromite
HYDROUS OXIDS
  Diaspore
  Gibbsite
  Cliachite
  Brucite
  Limonite
CARBONATES
  { Calcite
  { Dolomite
  { Magnesite
  { Siderite

CARBONATES (*Continued*)
  Aragonite
SULFATES
  { Barite
  { Celestite
  Anhydrite
  Gypsum
  Polyhalite
  { Alunite
  { Jarosite
PHOSPHATES
  Monazite
  { Apatite
  { Dahllite
  Collophane
  Lazulite
TITANATES
  Perovskite
SILICATES
  THE FELDSPARS
    Orthoclase
    Adularia
    Sanidine
    Microcline
    Anorthoclase
    { Albite
    { Oligoclase
    { Andesine
    { Labradorite
    { Bytownite
    { Anorthite
  THE FELDSPATHOIDS
    Leucite
    Nepheline
    Cancrinite
    { Sodalite
    { Haüyne
    Melilite

CLASSIFICATION OF THE MINERALS AND MINERALOIDS DESCRIBED (*Continued*)

SILICATES (*Continued*)

THE PYROXENES
{ Enstatite
{ Hypersthene
  Diopside
  Augite
  Pigeonite
  Hedenbergite
  Aegirine-augite
  Aegirine
  Jadeite
  Spodumene

THE PYROXENOIDS
  Wollastonite

THE AMPHIBOLES
  Anthophyllite
{ Cummingtonite
{ Grunerite

  Tremolite—Actinolite

  Nephrite
  Hornblende
  Lamprobolite
  Riebeckite
  Glaucophane

Olivine Group
  Forsterite
  Olivine
  Fayalite
  Monticellite
Chondrodite

Garnet Group
Beryl
Scapolite Group
Idocrase
Zircon
Topaz
Andalusite

Sillimanite
Kyanite
Mullite
Dumortierite

Tourmaline Group
  Schorlite
  Dravite
  Elbaite
Axinite

Epidote Group
  Zoisite
{ Clinozoisite
{ Epidote
{ Piedmontite
{ Allanite
Staurolite
Sphene
Cordierite
Prehnite
Lawsonite

Mica Group
{ Muscovite
{ Lepidolite
{ Phlogopite
{ Biotite

Chlorite Group
{ Clinochlore
{ Pennine
{ Prochlorite
  Chamosite
Chloritoid
Talc
Pyrophyllite

Clay Minerals
  Kaolinite
  Dickite
  Halloysite
  Montmorillonite

CLASSIFICATION OF THE MINERALS AND MINERALOIDS DESCRIBED (*Continued*)

SILICATES (*Continued*)
  Hydromuscovite
  SERPENTINE MINERALS
    ANTIGORITE
    Serpophite
    Chrysotile
  Iddingsite
  Glauconite
  THE ZEOLITES
    Analcime
    Heulandite

SILICATES (*Continued*)
    Stilbite
    Chabazite
    Natrolite
    Mesolite
    Thomsonite
    Scolecite
  THE LESS DEFINITE MINERAL-
    OIDS
    VOLCANIC GLASS
    Palagonite

GRAPHITE

C   Opaque   Hexagonal

Color.—Black with metallic luster in reflected light. It is opaque even in the thinnest sections.

Form.—The characteristic occurrence of graphite is in thin flakes and disseminated scales. The crystals are tabular in habit.

Distinguishing Features.— In thin sections graphite resembles magnetite, but its flaky appearance is distinctive for an opaque mineral.

Occurrence.—Graphite is characteristic of metamorphic rocks such as schists, gneisses,

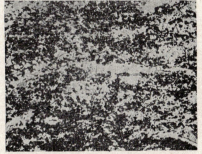

FIG. 137.—(×20) Graphite in schist.

slates, and metamorphic limestones. It gives the gray color to many metamorphic limestones and in some cases is the only mineral present in addition to calcite.

CARBONACEOUS MATTER

It is very common to find, in some rocks, finely divided opaque particles that cannot be referred to any definite mineral. This material is usually carbonaceous matter that has not been subjected to a temperature high enough to produce graphite.

Sphalerite

(Zn,Fe)S   Isometric

$$n = 2.37 \text{ to } 2.47$$

(Increases with iron content)

Color.—Gray to yellow or brown in thin sections; the color is not always uniform.

Form.—Sphalerite is usually found in anhedral crystals.

Cleavage.—More or less prominent parallel to {110}.

Relief.—Very high, $n >$ balsam; adamantine luster by reflected light.

Birefringence nil. Dark between crossed nicols, but it may be difficult to recognize the isotropic character on account of total reflection.

Distinguishing Features.—The very high relief, isotropic character, and cleavage are characteristic.

Occurrence.—Sphalerite is a common and widely distributed mineral in veins and replacement deposits. The common associates are pyrite, galena, marcasite, and the gangue minerals, quartz, chalcedony, calcite, dolomite, and siderite. It occurs in a few sedimentary rocks but is rare as a rock-forming mineral.

PYRITE

FeS$_2$ Opaque   Isometric

Pyrite is the most common sulfid mineral.

Fig. 138.—($\times 28$) Euhedral pyrite crystals with phlogopite (gray) in schist-gneiss.

Color.—Brass colored with metallic luster in reflected light. Opaque even in the thinnest sections.

Form.—Pyrite is common in euhedral crystals, which are usually cubes, showing in thin sections square, rectangular, or triangular outlines. It also occurs in irregular grains, masses, and veinlets.

Distinguishing Features.— Pyrite resembles pyrrhotite and chalcopyrite but is distinguished by color differences in reflected light. Chalcopyrite is deeper yellow. Pyrrhotite is bronze colored. Pyrite is even more like marcasite, and it may be necessary to use polished surfaces to differentiate the two minerals.

Alteration.—Pyrite is often more or less altered to limonite and occasionally to an indefinite hydrous iron oxid with a red streak.

Occurrence.—Pyrite is a very common and widely distributed mineral in veins and a secondary mineral in igneous rocks. It is also found in many sedimentary and metamorphic rocks.

Pyrrhotite

FeS(S)$_x$ Opaque   Hexagonal

Color.—Bronze colored with metallic luster in reflected light. Opaque even in the thinnest sections.

Form.—Pyrrhotite usually occurs in grains and irregular masses. Euhedral crystals are exceedingly rare.

Cleavage.—Parting parallel to {0001} is fairly common in pyrrhotite. In consequence it often shows a platy structure.

Distinguishing Features.— Pyrrhotite resembles pyrite but is bronze rather than brass colored. It is scratched by a knife blade while pyrite is not. Pentlandite, a rare

FIG. 139.—(×28) Pyrrhotite with phlogopite in metamorphic limestone.

nickel-iron sulfid, is a common associate of pyrrhotite and very much resembles it, but pentlandite has a characteristic octahedral parting. This does not always show, and it may be necessary to use polished surfaces to distinguish the two minerals.

Occurrence.—Pyrrhotite occurs in igneous rocks as a late magmatic mineral. It is also found in veins and in some metamorphic rocks. Compared with pyrite it is a high-temperature mineral.

Chalcopyrite

CuFeS$_2$ Opaque   Tetragonal

Chalcopyrite is the most widely distributed of the numerous copper minerals.

Color.—Brass colored with metallic luster in reflected light. Opaque even in the thinnest sections.

Form.—Chalcopyrite rarely occurs in euhedral crystals. It is found in anhedra and in veinlets.

Distinguishing Features.—Chalcopyrite resembles pyrite but is a deeper yellow. It is scratched by a knife; pyrite is not.

Occurrence.—Chalcopyrite is a very common mineral in veins and ore deposits. It is occasionally found in igneous, sedimentary, and metamorphic rocks.

Halite

NaCl   Isometric

$$n = 1.544$$

Color.—Colorless but may contain inclusions.

Form.—Halite is not found in thin sections prepared in the ordinary way, but the sections may be ground in glycol. The halite usually appears in anhedral crystals.

Cleavage perfect cubic.

Relief very low, n being about the same as balsam or clove oil.

Birefringence nil. Dark between crossed nicols.

Distinguishing Features.—The very low relief, cubic cleavage, and solubility are characteristic. About the only mineral that is likely to be mistaken for halite is sylvite, but the latter has appreciable relief and an index less than balsam (for sylvite $n = 1.490$).

Occurrence.—Halite occurs in sedimentary beds of rock salt that are often accompanied by anhydrite and gypsum. Sylvite and polyhalite are characteristic associates.

Fluorite

CaF$_2$ Isometric

$$n = 1.434$$

Color.—Colorless or purple in bands or spots.

Form.—Fluorite is sometimes found in euhedral crystals with square outline, but it is usually anhedral and often fills the spaces between other minerals.

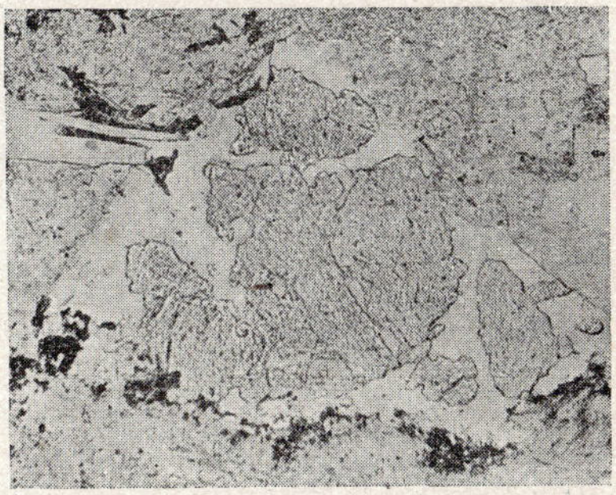

Cleavage perfect octahedral {111}. The cleavage usually appears as two intersecting lines at oblique angles of 70 and 110°, occasionally at three intersecting lines of 60 and 120°.

FIG. 140.—($\times$26) Fluorite with quartz in granite.

Relief fairly high, $n <$ balsam. According to Merwin, the index of refraction is remarkably constant (1.4338 for sodium light). Dispersion of fluorite is very low; hence the use of fluorite for apochromatic objectives.

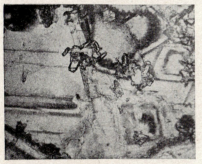

FIG. 141.—($\times$9) Fluorite showing zonal structure.

Birefringence nil. Dark between crossed nicols.

Distinguishing Features.— The rather high relief, perfect cleavage, and isotropic character distinguish fluorite from practically all other minerals. The purple spots or bands (see Fig. 141) are very characteristic.

Occurrence.—Fluorite is a common vein mineral, but it is rather rare in rocks in general. It is found in some granites, occasionally in sandstones, limestones, and phosphorites.

THE SILICA MINERALS AND MINERALOIDS

Silica occurs in nature in the six distinct minerals and mineraloids listed on this page.   Of these the first three are very common.   Lechatelierite (silica glass) is exceedingly rare.   The other two, tridymite and cristobalite, are widely distributed in volcanic rocks and can hardly be called rare minerals.   They have frequently been overlooked.

In respect to physical properties the silica minerals may be placed in two groups.   Quartz and chalcedony have refractive

THE SILICA MINERALS AND MINERALOIDS

Mineral	Crystal system	Indices of refraction
Quartz.........	Hexagonal	$n_\omega = 1.5442$, $n_\epsilon = 1.5533$, $n_\epsilon - n_\omega = 0.009$
Chalcedony....	Aggregates	$n_\alpha = 1.531$, $n_\gamma = 1.539$, $n_\gamma - n_\alpha = 0.008$
Opal..........	Amorphous	$n = 1.40$–1.46
Tridymite.....	Pseudo-hexagonal	$n_\alpha = 1.469$, $n_\gamma = 1.473$, $n_\gamma - n_\alpha = 0.004$
Cristobalite....	Pseudo-isometric	$n_\alpha = 1.484$, $n_\gamma = 1.487$, $n_\gamma - n_\alpha = 0.003$
Lechatelierite..	Amorphous	$n = 1.458$–1.462

indices near that of balsam and birefringence of about 0.009; the other four have lower indices of refraction and weaker birefringence, which reaches nil in lechatelierite and usually in opal.

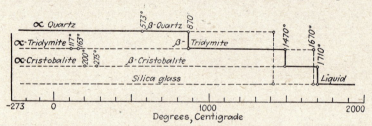

FIG. 142.—Diagram to show the relations between the various forms of silica.
(*After Sosman.*)

Chalcedony always occurs in aggregates of some sort, and consequently the optical properties are not completely known.

On heating quartz there is a sudden change in the properties at 573°C.   The symmetry changes from $A_3.3A_2$ to $A_6.6A_2$.   The low-temperature form is called α-*quartz* or *low quartz* and the high-temperature form, β-*quartz* or *high quartz*.   The quartz of igneous rocks was β-quartz at the time of its formation but on cooling inverted to α-quartz.   Similar changes take place on heating tridymite and cristobalite.   The stability range of the various silica minerals is shown in the diagram of Fig. 142.

Chalcedony and opal are low-temperature minerals.   Quartz has a considerable temperature range.   Tridymite and cristobalite also have a rather large temperature range.   The two latter occur almost exclusively in volcanic igneous rocks and in all probability have been formed by hot gases at the close of the magmatic period.

α-QUARTZ

SiO$_2$ (Low Quartz)   Hexagonal
(Rhombohedral Subsystem)

$$n_\omega = 1.5442$$
$$n_\epsilon = 1.5533$$
$$\text{Opt. } (+)$$

Color.—Colorless in thin sections.   It often contains inclusions.

Form.—Quartz occurs in euhedral prismatic crystals, in veinlets, disseminated grains, and as replacement anhedra.   It may be intergrown with orthoclase or microcline (graphic granite) and with plagioclase in vermicular forms (myrmekite).   It often occurs as a late interstitial mineral.   Quartz is common as pseudomorphs after other minerals.

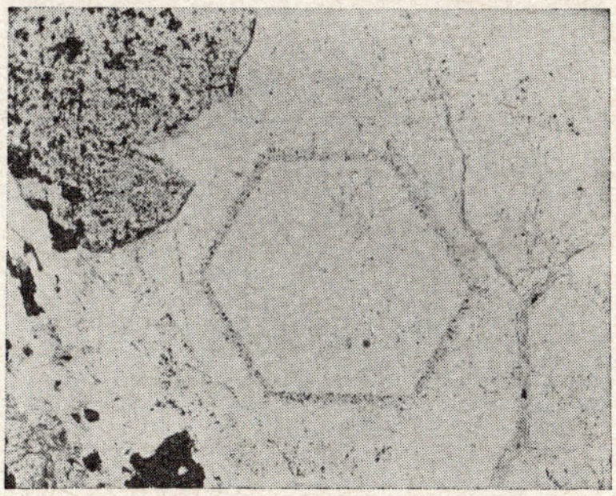

Fig. 143.—($\times 25$) Quartz (α-quartz) showing growth stages marked by inclusions.

Cleavage usually absent, but it sometimes shows on the edge of the slide.   The cleavage is imperfect rhombohedral $\{10\bar{1}1\}$, almost rectangular in favorable sections since $rr' = 85°46'$.

Relief very low, $n >$ balsam.

Birefringence rather weak, $n_\epsilon - n_\omega = 0.009$; thin sections 0.03 mm. thick show as a maximum, first-order white interference color with a slight tinge of yellow. Quartz is very useful in determining the thickness of any slide in which it occurs.

Extinction parallel in euhedral crystals and symmetrical to cleavage traces. Basal sections are dark in all positions. Irregular and wavy extinction due to strain is common. Vein quartz often shows peculiar structures such as flamboyant, feathered, lamellar, etc. Secondary enlargements of quartz grains are common in sandstones and quartzites.

Fig. 144.—(×9) Vein quartz (α-quartz). (× nicols.)

Orientation.—The position of the slower ray marks the trace of the *c*-axis. Euhedral crystals are therefore length-slow.

Twinning.—Although twinning is common in quartz, it rarely shows in thin sections.

Interference Figure.—Basal sections of ordinary thickness give a uniaxial positive figure without any rings. The interference figure of thick sections (greater than 1 mm.) has a weak or hollow center on account of rotary polarization.

Occasionally quartz gives a biaxial figure with 2V as high as 10°.

Distinguishing Features.— Quartz is usually easy to determine on account of its lack of alteration, absence of cleavage except perhaps on the edge of the section, and

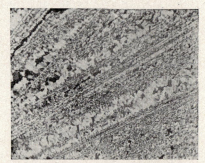

Fig. 145.—(×11) Quartz (α-quartz) pseudomorphous after lamellar calcite. (× nicols.)

absence of twinning. Cordierite may be mistaken for it, but cordierite is biaxial. Beryl resembles quartz in thin sections but is length-fast and optically negative. Some varieties of scapolite also resemble quartz, but they are optically negative, length-fast, and have cleavage. Chalcedony has aggregate structure.

Alteration.—Quartz is less affected by alteration than almost any other mineral, but it sometimes shows slight replacement by sericite, by pyrophyllite, and rarely by talc.

Occurrence.—Quartz is a ubiquitous mineral. It is found in many rock types as an essential, accessory, or secondary mineral. It is especially abundant in sandstones, arkoses, sands, quartzites, granites, rhyolites, and gneisses. In many igneous rocks it is a secondary mineral in seams and cavities. Quartz is the most common of all vein minerals. It occurs as a replacement of other minerals and as a replacement of wood and calcareous fossils.

Quartz is one of the most common detrital minerals.

β-QUARTZ

SiO₂ (High Quartz) Hexagonal
(Hexagonal Subsystem)

The quartz appearing as phenocrysts in rhyolites and quartz porphyries is the high-temperature form known as *β-quartz*. It has formed above 573° C.; and the low-temperature form called *α-quartz*, below 573° C. On cooling, the β-quartz inverts to the α-form, so that all quartz examined in thin sections is now α-quartz. The habit of β-quartz is usually different from that of the α-form. A hexagonal dipyramid (the symmetry is $A_6.6A_2$) predominates,

Fig. 146.—(×12) Graphic intergrowth of β-quartz and feldspar. (× nicols.)

and the prism face is subordinate, whereas in the α-form the prism predominates.

As shown by Drugman, the twinning laws of β-quartz are different from those of α-quartz, but twinning rarely shows in thin sections.

CHALCEDONY

SiO₂

$$n_\alpha = 1.531$$
$$n_\gamma = 1.539$$

Color.—Colorless to pale brown in thin sections and often bluish white by reflected light.

Form.—Chalcedony usually occurs as a cavity filling or lining that is often spherulitic (see Fig. 148), as a replacement of fossils, as cementing material, and in massive form.

FIG. 147.—(×15) Banded chalcedony and quartz in veinlets in jasper.

Relief low, n about the same as that of balsam, either slightly lower or slightly greater.

Birefringence rather weak, $n_\gamma - n_\alpha = 0.008$, practically the same as that of quartz. Chalcedony always shows aggregate structure between crossed nicols. This often takes on a spherulitic form with the spherulitic cross prominent in many cases.

Extinction parallel to the length of the fibers.

Orientation.—The fibers are usually length-fast, but in many cases they are length-slow. The fibers of concentric zones are often alternately slow and fast.

Distinguishing Features.— The aggregate structure with optical properties very close to

FIG. 148.—(×45) Spherulitic chalcedony. (× nicols.)

those of quartz is distinctive for chalcedony. The minerals most likely to be mistaken for chalcedony are probably gibbsite and dahllite, but in both of these the relief in balsam or clove oil is distinctly higher.

Occurrence.—Chalcedony is a secondary mineral in the cavities of igneous rocks and is often associated with quartz, opal, and the zeolites. It also occurs in sedimentary limestone in nodules and bands and as a replacement of calcareous fossils. Chalcedony is the principal constituent of cherts and jaspers. It occurs in diatomite as a replacement of opal. The temperature range of chalcedony seems to be lower than that of quartz.

OPAL

$SiO_2(H_2O)_x$ Amorphous
(Mineraloid)

$$n = 1.40 \text{ to } 1.46$$
(usually *ca.* 1.45)

Like other mineraloids opal is variable in its properties.

Color.—Colorless to pale gray or brown in thin sections.

Form.—Opal is often found in colloform crusts, in veinlets, and as a cavity filling or lining. More often it is massive without any particular structure. It often occurs as a replacement of wood and other organic materials. It is common as a replacement of feldspar and as the cementing material in sandstone.

Cleavage absent, but irregular fractures are found on the edges of thin sections.

Relief rather high, but n < balsam.

Fig. 149.—(×34) Colloform opal with infilled chalcedony.

Birefringence usually nil, but some varieties, especially hyalite, may show very weak birefringence that is due to strain. Interference colors caused by exceedingly thin films show in sections of precious opal, especially in reflected light.

Distinguishing Features.—The high relief and low index of refraction are distinctive. Lechatelierite (silica glass) is very similar, and it may be necessary to try the closed-tube test for water in order to distinguish them.

Occurrence.—Opal is a secondary mineraloid in volcanic igneous rocks. It appears in cavities or seams and as a replacement of feldspars or other silicates. The more common associates are quartz, chalcedony, and tridymite. It is the principal constituent of diatomite and geyserite and occasionally occurs as the cementing material in sandstone. It also occurs as the main constituent of opal shale and opal rock.

A fibrous variety of silica with a low index of refraction known as *lussatite* has been considered to be a variety of tridymite or even a distinct silica mineral, but it is probably a mixture of fibrous chalcedony and opal.

Tridymite

SiO$_2$ Orthorhombic
(Hexagonal above 117° C.)

$$n_\alpha = 1.469$$
$$n_\beta = 1.469$$
$$n_\gamma = 1.473$$
$$2V = 35°; \text{Opt. } (+)$$

Color.—Colorless in thin sections.

Form.—Tridymite usually occurs in minute euhedral crystals as a cavity lining. The crystals are six sided, thin tabular, and are often twinned. It also occurs as a porous crystalline aggregate.

Relief moderate, but $n <$ balsam.

Birefringence very weak, $n_\gamma - n_\alpha = 0.004$; best seen with a sensitive-violet test plate.

Twinning.—Wedge-shaped twins made up of two or three individuals are characteristic. The twin plane is {10$\bar{1}$6}.

Interference Figure.—Because of the small size of the crystals it is very difficult to obtain interference figures with tridymite.

Distinguishing Features.—Tridymite very much resembles cristobalite, not only in its general appearance but also in its geologic occurrence. (In the experience of the senior author, the tile structure often said to be characteristic of this mineral is not at all common.) The twinning of tridymite and the wedge-shaped sections are characteristic; in their absence it may be necessary to determine the refractive index of isolated grains. (For tridymite $n < 1.480$; for cristobalite $n > 1.480$.)

Alteration.—Paramorphs of cristobalite after tridymite have been found at several localities by the senior author. Paramorphs of quartz after tridymite (the pseudo-tridymite of Mallard) are known from a number of localities.

Occurrence.—The characteristic occurrence of tridymite is in the cavities of volcanic igneous rocks such as obsidian, rhyolite, andesite, etc. It is a late mineral formed by hot gases. Tridymite is not very abundant, but it is common and widely distributed. A tridymite-feldspar rock formed by the action of hot gases upon rhyolitic obsidian has been

Fig. 150.—(×50) Twinned tridymite in cavity of altered rhyolitic glass. (The central wedge of tridymite is in the extinction position.) (× nicols.)

studied by the senior author in Imperial County, California. In Texas tridymite occurs in a rhyolitic tuff (Gueydan formation of the southwestern Gulf Coastal Plain).

Artificial Tridymite

The principal constituent of silica brick is tridymite with cristobalite as an associate. The silica bricks are made by heating ground-up quartzites of low iron content. The best bricks are said to be those with the largest amount of tridymite.

Cristobalite

SiO_2 Pseudo-isometric

(Isometric above 230° C.)

$$n_\alpha = 1.484$$
$$n_\gamma = 1.487$$

Color.—Colorless in thin sections.

Form.—Cristobalite is found in minute square crystals or aggregates in the cavities of volcanic igneous rocks; it also occurs intergrown with the feldspar fibers of spherulites (see Fig. 202).

Cleavage.—Cristobalite has a peculiar curved fracture that is highly characteristic.

Relief moderate, $n <$ balsam.

Birefringence very weak, $n_\gamma - n_\alpha = 0.003$; best detected with a sensitive-violet test plate. Between crossed nicols it often shows a mosaic structure that is due in part to twinning.

Distinguishing Features.—Cristobalite closely resembles tridymite, but the curved fracture usually distinguishes it. It may

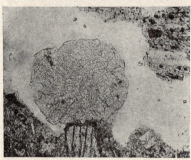

FIG. 151.—($\times$20) Cristobalite in a cavity in auganite.

be necessary to determine the refractive index of detached fragments. (For cristobalite $n > 1.480$; for tridymite $n < 1.480$.)

Occurrence.—Cristobalite is found in volcanic igneous rocks such as obsidian, rhyolite, andesite, auganite, and basalt. The fact that it usually occurs in cavities is evidence that it has been formed by hot gases at a late stage. Tridymite is a common associate, and at a number of localities paramorphs of cristobalite after tridymite have been noted.

Cristobalite has been found in some specimens of opal and hence has been formed at a temperature lower than its stability range.

ARTIFICIAL CRISTOBALITE

Cristobalite and tridymite are the constituents of silica bricks that are made by heating ground-up quartzites of low iron content.

Lechatelierite

SiO_2 (Silica Glass)   Amorphous
(Mineraloid)

$$n = 1.458 \text{ to } 1.462$$

Color.—Colorless in thin sections. The tendency toward opacity is due to minute bubbles.

Form.—Lechatelierite is amorphous silica glass. It is usually vesicular and may also be banded and show flow structure.

Relief low, $n <$ balsam.

Birefringence nil. Dark between crossed nicols.

Distinguishing Features.—From other glasses lechatelierite may be distinguished by its very low refractive index. It very much resembles opal except in its geologic occurrence. A closed-tube test may be necessary to make certain that the mineral is not opal.

Occurrence.—Lechatelierite is the main constituent of fulgurites, which are hollow tubes of glass produced by the action of lightning upon quartzose sand.

An interesting occurrence of lechatelierite is that of Meteor Crater, Arizona. Here a highly vesicular silica glass

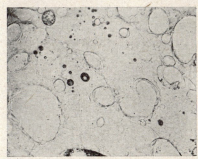

Fig. 152.—(×95) Lechatelierite (silica glass) with air vesicles. (Small dark areas are bubbles in the Canada balsam.)

has been produced from sandstone by the heat generated as a result of the explosive impact of a huge meteorite or meteorite swarm (Cañon Diablo meteorite).

Artificial Lechatelierite

Silica glass is now made artificially on a large scale for various kinds of chemical apparatus, lenses, and window panes to transmit ultra-violet light. It has a remarkably low coefficient of thermal expansion.

Periclase

MgO   Isometric

$$n = 1.738 \text{ to } 1.760$$

Periclase is a rare but widely distributed mineral.

Color.—Colorless in thin sections.

Form.—Periclase occurs in equant crystals or anhedral crystal aggregates. Individual anhedra may be recognized by cleavage traces.

Cleavage cubic. Parting, dodecahedral.

Relief high, $n >$ balsam.

Birefringence nil. Dark between crossed nicols.

Distinguishing Features.—The cubic cleavage, high relief, and isotropic character taken together are distinctive.

Alteration.—Periclase is usually altered to brucite, which, in turn, may be altered to hydromagnesite.

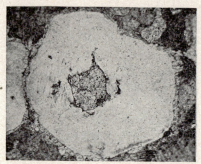

FIG. 153.—(×13) Core of periclase surrounded by brucite in metamorphic limestone.

Occurrence.—The most common occurrence of periclase is in metamorphic limestones. It is found as cores within brucite spots that are formed by hydration of the periclase.

ARTIFICIAL PERICLASE

Artificial periclase, now prepared on a commercial scale from selected magnesite, is more familiar than the natural mineral. It is used as a high-grade electric insulator.

Corundum

Al_2O_3

Hexagonal
(Rhombohedral Subsystem)

$$n_\epsilon = 1.759 \text{ to } 1.763$$
$$n_\omega = 1.767 \text{ to } 1.772$$
$$\text{Opt. } (-)$$

Color.—Usually colorless, sometimes with blue or pink areas that are not uniformly colored. Zoned crystals are not uncommon. In thick sections corundum may be pleochroic.

Form.—Euhedral crystals are common. The habit varies from tabular to prismatic; cross sections are six sided and may show zonal structure. Skeleton crystals are often encountered.

FIG. 154.—(×22) Corundum crystal with rhombohedral parting in corundum syenite.

Cleavage.—Parting often parallel to the unit rhombohedron $\{10\bar{1}1\}$ or the pinacoid $\{0001\}$ or both.

Relief very high, $n >$ balsam.

Birefringence weak, $n_\omega - n_\epsilon = 0.008\text{--}0.009$. Sections are usually thicker than normal on account of the extreme hardness of the corundum. For this reason the maximum interference color of most thin sections runs up into the second order.

Extinction parallel to the crystal outlines or symmetrical to the rhombohedral parting.

Orientation.—Sections of tabular crystals are length-slow, and sections of prismatic crystals are length-fast since the optic sign of the mineral is negative.

Twinning.—Twinning lamellae or twin seams with $\{10\bar{1}1\}$ as the twin-plane are rather common.

Interference Figure.—The figure obtained in basal sections is uniaxial negative usually with one ring. Some figures are biaxial with 2V as high as 30°.

FIG. 155.—(×25) Corundum crystals of tabular habit in a ground mass of muscovite and pyrophyllite. (× nicols.)

Distinguishing Features.—The combination of very high relief with weak birefringence, parting, and twinning lamellae is distinctive.

Occurrence.—Corundum is especially characteristic of corundum syenites, contact-metamorphic limestones, and metamorphosed shales. It may also be found in schists and as a sporadic detrital mineral. It is one of the principal constituents of emery, which is probably a metamorphosed bauxite or laterite. In igneous rocks it never occurs with original quartz.

Hematite

Fe_2O_3 Hexagonal
                                        (Rhombohedral Subsystem)

$$n_\epsilon = 2.94$$
$$n_\omega = 3.22$$
$$\text{Opt. } (-)$$
$$\text{(Usually opaque)}$$

Color.—Black with metallic luster in reflected light.   Usually opaque, but some varieties are translucent red.

Form.—Hematite occurs in anhedral crystals, grains, masses, and occasionally minute scales. It is found as inclusions in other minerals.

Cleavage.—Hematite sometimes exhibits pseudo-cubic rhombohedral parting (parallel to $10\bar{1}1$).

Birefringence.—The translucent red scales are dark between crossed nicols since they are parallel to (0001).

Distinguishing Features.—Generally speaking, hematite resembles magnetite but is a somewhat different color by reflected light.   It is non-magnetic and is often translucent red.

Occurrence.—Hematite is rare as an original constituent of igneous rocks, but it is rather common as a secondary mineral in many types of rocks. It occurs as the main constituent of hematite schist.

ILMENITE

$FeTiO_3$ Opaque   Hexagonal
                                        (Rhombohedral Subsystem)

Color.—Black with metallic luster.   It is opaque even in the thinnest sections.

Form.—Ilmenite occurs in disseminated tabular crystals, which usually appear as elongate sections, in irregular grains, and sometimes in masses.   It may also appear as an intergrowth with magnetite, but polished surfaces are usually necessary to reveal it.

Distinguishing Features.—Ilmenite is easily mistaken for magnetite or hematite but may as a rule be recognized by the tendency to form skeleton crystals and by the white alteration product (leucoxene).

Alteration.—Ilmenite is usually more or less altered to an opaque white material called *leucoxene*.

Occurrence.—Ilmenite is a widely distributed mineral in some types of igneous rocks, more especially diabases and dolerites. It also occurs in some kinds of refractory iron ore. Ilmenite is an important constituent of many "black sands."

FIG. 156.—(×12) Skeleton crystal of ilmenite.

LEUCOXENE

An opaque white substance called *leucoxene* is common as an alteration product in various rocks. It occurs on the surface and around the borders of ilmenite and is also disseminated through various rocks and is probably the result of hydrothermal alteration. Leucoxene has sometimes been identified with sphene. Leucoxene in a detrital deposit in Oklahoma is amorphous hydrous titanium dioxid, according to Coil.

Rutile

TiO_2 Tetragonal

$$n_\omega = 2.603 \text{ to } 2.616$$
$$n_\epsilon = 2.889 \text{ to } 2.903$$
$$\text{Opt. } (+)$$

Color yellowish to reddish brown in thin sections. In reflected light it shows adamantine luster.

Form.—Rutile usually occurs in small prismatic to acicular crystals and in grains. Knee-shaped twins with {101} as the twin-plane are characteristic. Capillary crystals are common, especially in quartz.

Cleavage parallel to the length of the crystals {110}.

Relief very high, $n >$ balsam. Adamantine luster by reflected light.

Birefringence extreme, $n_\epsilon - n_\omega = 0.286-0.287$; interference colors are very high but do not show well on account of total reflection.

Extinction parallel.

Twinning common (see under Form).

Distinguishing Features.—The mineral most likely to be mistaken for rutile is probably baddeleyite (ZrO_2), which sometimes occurs in corundum syenites. The color, together with very high relief, is distinctive.

Related Minerals.—Two polymorphs of rutile, anatase (also tetragonal) and brookite (orthorhombic) are of importance as detrital minerals.

Occurrence.—Rutile is a rather widely distributed accessory mineral in various metamorphic rocks. It occasionally occurs in igneous rocks such as the albitite of Kragerö, Norway. Rutile also occurs as a detrital mineral. Sphene is a common associate.

Cassiterite

SnO_2 Tetragonal

$$n_\omega = 1.996$$
$$n_\epsilon = 2.093$$
$$\text{Opt. } (+)$$

Color.—Colorless to gray, yellowish, reddish, or brown in thin

sections. It often shows zones of varying color.

Form.—Cassiterite is usually found in subhedral crystals. Veinlets are rather common.

Cleavage prismatic, parallel to the length.

Relief very high, $n >$ balsam. Adamantine luster in reflected light.

Birefringence extreme,

FIG. 157.—($\times 16$) Cassiterite with quartz.

$n_\epsilon - n_\omega = 0.097$; the interference colors are high order but are usually masked by the color of the mineral.

Extinction parallel to the cleavage, oblique to the twin-plane.

Twinning.—Twinned crystals are common; the twin-plane is $\{101\}$.

Distinguishing Features.—Cassiterite resembles sphalerite, but the latter is isotropic. From rutile it is distinguished by lower birefringence.

Occurrence.—Cassiterite occurs in granite pegmatites, in greisen, and in high-temperature veins. The usual associates are quartz, muscovite, schorlite, and topaz. The wood-tin variety occurs in rhyolites.

THE SPINEL GROUP

The spinels are aluminates, ferrites, and chromites of dyad metals magnesium and iron. They occur in isometric crystals, usually octahedrons, and are optically isotropic.

In addition to the common spinels described here—spinel, magnetite, and chromite—there are hercynite ($FeAl_2O_4$), magnesioferrite, ($MgFe_2O_4$), galaxite ($MnAl_2O_4$), franklinite [$(Zn,Mn)Fe_2O_4$], and gahnite ($ZnAl_2O_4$).

Spinel

$(Mg,Fe)(Al,Cr)_2O_4$ Isometric
$$n = 1.72 \text{ to } 1.78$$

Color.—Colorless to green (pleonaste), olive green, or brown (picotite) in thin sections.

Form.—Spinel practically always occurs in euhedral or subhedral crystals or in equant grains. Crystals are octahedra, and the most common sections are rhombic in outline.

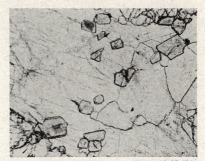

Cleavage imperfect octahedral, but it may not show.

Relief high, $n >$ balsam.

Birefringence nil. It is one of the few isometric minerals that is invariably isotropic.

Twinning.—Twinning according to the spinel law with {111} as twin-plane is rather common, but it does not usually show in the slide.

FIG. 158.—($\times 18$) Spinel (high relief) in metamorphic limestone.

Distinguishing Features.—Pleonaste, iron-bearing variety of spinel, is much like hercynite.

Spinel is distinguished from garnet by its octahedral form.

Related Minerals.—A related mineral, hercynite $FeAl_2O_4$, is a prominent constituent of certain types of emery. Picotite is intermediate between spinel and chromite. It resembles chromite but is more transparent. Galaxite is manganese spinel with the formula $MnAl_2O_4$. It occurs with alleghanyite.

Occurrence.—Spinel occurs in metamorphic limestone with phlogopite and chondrodite, in other metamorphic rocks, and also in various igneous rocks. It is rare as a detrital mineral.

Picotite is common in peridotites, dunites, and derived serpentines.

MAGNETITE

$Fe^{II}Fe_2^{III}O_4$ Opaque   Isometric

Color.—Black with metallic luster in reflected light. Opaque even in the thinnest sections. Strongly magnetic with an ordinary magnet.

Fig. 159.—(×9) Euhedral to anhedral magnetite in norite.

Fig. 160.—(×12) Secondary magnetite in antigorite formed at the expense of olivine.

Form.—Magnetite is found in distinct crystals, irregular grains, and masses. The common crystal form is the octahedron, sections of which are triangles, squares, and rhombs. Skeleton crystals are sometimes encountered. Magnetite is common as an intergrowth with ilmenite.

Cleavage.—Octahedral parting is sometimes present.

Distinguishing Features.—The minerals most likely to be mistaken for magnetite are ilmenite, hematite, and chromite. Ilmenite usually occurs in skeleton crystals and often has an opaque white alteration product (leucoxene) on its surface or borders. Hematite is a different color from magnetite and is non-magnetic. Chromite is translucent brown on thin edges.

Occurrence.—Magnetite is a very common and widely distributed mineral in nearly all igneous and metamorphic rocks. In igneous rocks it is usually a late magmatic mineral (see Fig. 159) but sometimes occurs as a secondary one (see Fig. 160).

It is one of the most common detrital minerals and is one of the chief constituents of many "black sands."

Chromite

$(Fe,Mg)(Cr,Al,Fe)_2O_4$ Isometric

$$n = 2.07 \text{ to } 2.16$$

The chromite of meteorites approaches the composition $FeCr_2O_4$, but terrestrial chromite always contains appreciable amounts of magnesium and aluminum.

Color.—Black with submetallic luster in reflected light. Opaque in general in thin sections, but thin edges show translucent brown.

Form.—Chromite usually occurs in subhedral crystals, grains, or aggregates. Occasionally it is found in minute octahedra.

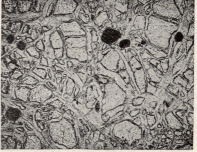

FIG. 161.—(×39) Chromite in altered dunite, showing olivine and antigorite.

Birefringence nil. The translucent brown edges are dark between crossed nicols.

Distinguishing Features.—Chromite is easily mistaken for magnetite, but the thin edges are translucent brown. If the section is a thick one, it may be well to crush some of the rock and make a rough concentrate. Minute particles of the chromite will appear translucent brown and isotropic.

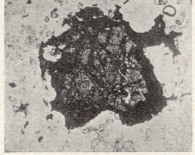

FIG. 162.—(×31) Chromite border around picotite and formed at its expense in serpentine.

Occurrence.—Chromite occurs for the most part in peridotites, pyroxenites, dunites, and derived serpentines. According to researches of the senior author, it is a late magmatic mineral in igneous rocks.

In most serpentines it is a relict mineral but may at times be formed during serpentinization at the expense of picotite.

Diaspore

$Al_2O_3.H_2O$　　　　　　　　　　　　　　　　　Orthorhombic

$$n_\alpha = 1.702$$
$$n_\beta = 1.722$$
$$n_\gamma = 1.750$$
$$2V = 84°; \text{ Opt. } (+)$$
$$a = \gamma \text{ or } Z, \; b = \beta \text{ or } Y, \; c = \alpha \text{ or } X$$

Color.—Colorless to pale blue in thin sections. Sometimes pleochroic in thick sections.

Form.—Diaspore occurs in tabular crystals parallel to {010}. The crystals may be rather large to very minute.

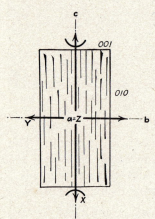

Cleavage perfect in one direction {010}.

Relief high, $n >$ balsam.

Birefringence strong, $n_\gamma - n_\alpha = 0.048$; so the maximum interference color is about upper third order.

Extinction parallel.

Orientation.—The crystals are length-fast.

Interference Figure.—The figure is biaxial positive with a very large axial angle. The axial plane is {010}. Dispersion, $r < v$ weak.

FIG. 163.—Orientation diagram of diaspore. Section parallel to (100).

Distinguishing Features.—Diaspore resembles andalusite and sillimanite but has much stronger birefringence.

Related Minerals.—Boehmite, dimorphous with diaspore, is found in some bauxites in very minute crystals.

Occurrence.—Diaspore occurs in metamorphic rocks such as schists and emery. It occurs in a few altered igneous rocks associated with alunite. It is also a prominent constituent of the highly aluminous flint clays.

Gibbsite

Al(OH)$_3$ (Hydrargillite)   Monoclinic
$\angle \beta = 85°29'$

$$n_\alpha = 1.554 \text{ to } 1.567$$
$$n_\beta = 1.554 \text{ to } 1.567$$
$$n_\gamma = 1.576 \text{ to } 1.589$$
$$2V = 0 \text{ to } 40°; \text{ Opt. } (+)$$
$$b = \alpha \text{ or } X, \; a \wedge \gamma \text{ or } Z = 25°$$

Color.—Colorless to pale brown in thin sections.

Form.—Gibbsite (called *hydrargillite* by European mineralogists) occurs in minute pseudo-hexagonal euhedral crystals in cavities and in fine crystalline aggregates that are often pseudomorphous after feldspars. Reticulate structure is common.

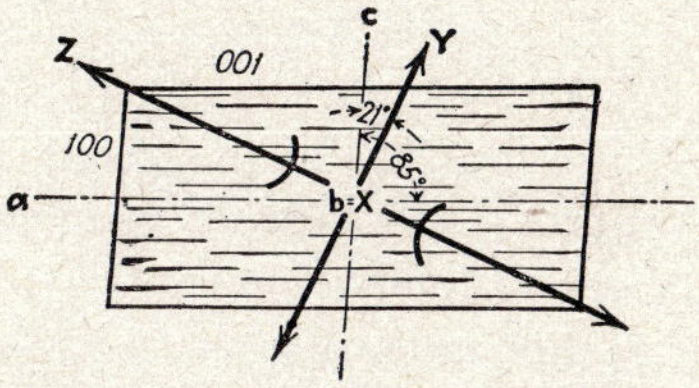

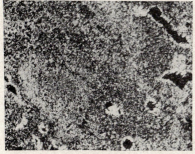

FIG. 164.—Orientation diagram of gibbsite. Section parallel to (010).

Cleavage in one direction parallel to {001}, but it may be difficult to see.

Relief moderate, $n >$ balsam.

Birefringence moderate, $n_\gamma - n_\alpha = 0.022$. The maximum interference colors are bright upper first-order or lower second-order colors.

Extinction.—Oblique extinction angles, up to a maximum of 25° for $a \wedge \gamma$ or Z in sections parallel to {010}.

Orientation.—Since the crystals are tabular parallel to {001}, elongate sections with twinning are length-slow.

FIG. 165.—(×25) Gibbsite in bauxite (microcystalline groundmass and minute crystals in cavities). (× nicols.)

Twinning.—Polysynthetic twinning with {001} as twin-plane is often sharp and well defined.

Interference Figure.—The crystals are usually too small to give an interference figure. In most crystals the axial plane is normal to (010), but in some it is parallel to (010).

Distinguishing Features.—On account of the aggregate structure gibbsite resembles chalcedony, but the relief is higher and the birefringence much stronger. It also resembles dahllite, but the latter mineral has weak birefringence.

Occurrence.—Some bauxites (bauxite is used as a rock name) are made up largely of gibbsite, others largely of amorphous cliachite with crystalline gibbsite in cavities.

Cliachite

$Al_2O_3(H_2O)_x$　　　　　(Bauxite in part)　　　　Amorphous
　　　　　　　　　　　　(Fig. 98, Page 116)　　　　(Mineraloid)
　　　　　　　　　　　$n = 1.57$ to 1.61

Color.—Colorless to deep brown or red in thin sections. Translucent to nearly opaque.

Form.—Cliachite is pisolitic or massive without any indication of crystalline structure.

Relief moderate, $n >$ balsam. It is difficult to test the relief and refractive index unless the mineral is powdered.

Birefringence nil. In favorable spots or in a powdered form the mineral is isotropic.

Distinguishing Features.—The pisolitic structure and association with gibbsite are distinctive.

Occurrence.—Cliachite is the main constituent of many bauxites. (Bauxite is here used as a rock name.) Common associates are gibbsite and siderite. There may also be relict minerals such as ilmenite and sphene, for in some cases bauxites are derived from nepheline syenites.

Bauxite

Since there is no known aluminum oxid dihydrate, either mineral or laboratory product, the term *bauxite* is used as a name for rocks consisting of aggregates of microcrystalline gibbsite or amorphous cliachite (a mineraloid), $Al_2O_3.(H_2O)_x$.

Brucite

Mg(OH)$_2$ Hexagonal
(Rhombohedral Subsystem)

$$n_\omega = 1.566$$
$$n_\epsilon = 1.585$$
$$\text{Opt. } (+)$$

Color.—Colorless in thin sections.

Form.—Brucite usually occurs in plates or scaly aggregates that appear fibrous in sections.

Cleavage perfect in one direction {0001} but may not show in thin sections.

Relief fair, $n >$ balsam.

Birefringence moderate, $n_\epsilon - n_\omega = 0.019$. Some of the interference colors are anomalous; a peculiar reddish brown hue takes the place of the yellow and orange of the first order. If the section is too thin, the anomalous colors do not show.

Fig. 166.—($\times$12) Massive brucite showing complicated texture. ($\times$ nicols.)

Extinction parallel.

Orientation.—The scaly aggregates, which are apparently fibrous, are length-fast.

Fig. 167.—($\times$18) Brucite showing intricate structure. ($\times$ nicols.)

Interference Figure.—The interference figure is uniaxial positive with the first ring anomalous (see under Birefringence). At times the figure may be biaxial with a small axial angle.

Distinguishing Features.—It resembles alunite but has better cleavage and anomalous interference colors.

Alteration.—Brucite is often altered to hydromagnesite, Mg$_4$(OH)$_2$(CO$_3$)$_3$.3H$_2$O.

Occurrence.—The most common occurrence of brucite is in metamorphic calcite-brucite rocks as an alteration of periclase, MgO. It is sometimes found in serpentine.

Limonite

$H_2Fe_2O_4(H_2O)_x$ Amorphous
                                                              (Mineraloid)

$$n = 2.0 \text{ to } 2.1$$

Although limonite is not a distinct mineral in the sense that hematite and magnetite are, it is a convenient term for the hydrous non-crystalline iron oxid with a yellow-brown streak.

Color.—Brown, translucent to opaque in thin sections.

Form.—Limonite usually occurs as a stain or border around other minerals. It sometimes appears in oolitic or pisolitic forms and as incrustations. Apparent crystals are pseudomorphs after pyrite.

Relief very high, $n >$ balsam.

Birefringence.—It is usually isotropic but may show irregular birefringence due to strain.

Related Minerals.—Goethite is similar to limonite, but it is definitely crystalline with parallel extinction.

Fig. 168.—(×12) Limonite cementing detrital fragments of quartz.

Occurrence.—Limonite is practically always a secondary mineraloid. It usually appears as an indefinite stain or coating but is sometimes a pseudomorph after other minerals, especially pyrite.

THE CALCITE GROUP

The calcite group of rhombohedral carbonates consists of the minerals listed below, together with rhodochrosite ($MnCO_3$)

CALCITE GROUP

Mineral	Chemical composition	n_ϵ	n_ω	$n_\omega - n_\epsilon$
Calcite.........	$CaCO_3$	1.486	1.658	0.172
Dolomite......	$Ca(Mg,Fe)(CO_3)_2$	1.500–1.526	1.680–1.716	0.180–0.190
Magnesite.....	$MgCO_3$	1.509–1.527	1.700–1.726	0.191–0.199
Siderite........	$FeCO_3$	1.596–1.633	1.830–1.875	0.234–0.242

and smithsonite ($ZnCO_3$). They are hexagonal (rhombohedral subsystem) with perfect rhombohedral cleavage and a cleavage angle of 73 to 75°. They are uniaxial and optically negative. All show change of relief when rotated; the higher relief is obtained when the long diagonal of the rhomb is parallel to the vibration plane of the lower nicol. The birefringence is extreme, and the maximum interference colors are high-order white.

Dolomite and magnesite may contain ferrous carbonate in isomorphous mixture, and this increases the value of the refractive indices.

Another member of the calcite group is rhodochrosite, $MnCO_3$. It is very similar to the other minerals in its optical properties. Rhodochrosite occurs in veins but is very rare as a rock-forming mineral. Still another member of the group is smithsonite. It usually occurs in the oxidized zone and is sometimes pseudomorphous after calcite and dolomite.

CALCITE

$CaCO_3$ Hexagonal
(Rhombohedral Subsystem)

$$n_\epsilon = 1.486$$
$$n_\omega = 1.658$$
$$\text{Opt. } (-)$$

Color.—Colorless in thin sections, but it is often cloudy.

Form.—Fine to coarse aggregates, usually anhedral. Euhedral crystals in rock sections are rare. Calcite often shows organic structure of some kind. It is frequently oolitic or spherulitic.

Cleavage perfect rhombohedral $\{10\bar{1}1\}$, usually shows at two intersecting lines at oblique angles (75° if section is cut normal to the cleavage traces). In fine aggregates the cleavage may not show. There is sometimes parting parallel to $\{01\bar{1}2\}$ which is due to twin-gliding.

Relief varies with the direction. It is high when the long diagonal of the rhomb is parallel to the vibration plane of the lower nicol and low when the short diagonal is in this position. Occasional sections parallel to $\{0001\}$ have high relief in all positions.

Birefringence extreme, $n_\omega - n_\epsilon = 0.172$. The maximum interference color is pearl gray or white of the higher orders.

Thin edges of the slide usually show bright colors of the fourth and fifth orders and tints of higher orders. Thin films and twin lamellae of calcite usually show bright interference colors.

Extinction symmetrical to the cleavage traces. When a section is in one of the extinction positions, fine birefringent calcite dust formed by grinding is prominent.

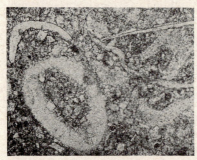

FIG. 169.—(×44) Calcite showing fragmentary fossils in sedimentary limestone.

Orientation is difficult to determine on account of the extreme birefringence.

Twinning.—Polysynthetic twinning with $\{01\bar{1}2\}$ as twin-plane is very common, especially in the calcite of metamorphic limestone. The twin lamellae are mostly parallel to the long diagonal, but they also intersect at oblique angles depending upon how the section is cut. The twin lamellae are usually so thin that they show first-order interference colors.

Interference Figure.—The interference figure is uniaxial negative with many rings. Cleavage flakes give a very eccentric figure. Occasionally calcite gives a biaxial figure with a small axial angle.

Distinguishing Features.—Dolomite, magnesite, and siderite may all be mistaken for calcite. Dolomite is usually subhedral to euhedral and often has twin lamellae parallel to the short diagonal as well as to the long diagonal. Siderite usually has iron stains

FIG. 170.—(×30) Anhedral crystals of calcite in metamorphic limestone.

around the borders of the grains, and the relief is not low in any position. Since there is no distinctive feature for magnesite, it may be necessary to make microchemical tests. Aragonite is also similar to calcite but lacks the rhombohedral cleavage, and in no section is the refractive index distinctly less than that of balsam. Aragonite is also biaxial.

Alteration.—Calcite is often replaced by quartz. A good example is shown in Fig. 145, page 186.

Occurrence.—Calcite is the principal constituent of both sedimentary and metamorphic limestones, but it is found in many other rock types. It is a very common secondary mineral in cavities of igneous rocks, where it is often associated with zeolites. It is also a deuteric mineral in some igneous rocks. Next to quartz, calcite is the most common vein mineral.

DOLOMITE

$Ca(Mg,Fe)(CO_3)_2$ (inc. Ankerite)   Hexagonal
(Rhombohedral Subsystem)

$$n_\epsilon = 1.500 \text{ to } 1.526$$
$$n_\omega = 1.680 \text{ to } 1.716$$
$$\text{Opt. } (-)$$

Color.—Colorless to gray.

Form.—Fine to coarse grained and usually subhedral. Euhedral crystals of the unit rhombohedron $\{10\bar{1}1\}$ are rather common, and the crystals are often curved. Zonal structure is frequent; this is due to variation in the iron content.

FIG. 171.—(×30) Euhedral crystals of dolomite in sedimentary dolomite.

Cleavage perfect rhombohedral parallel to $\{10\bar{1}1\}$, which usually shows as two intersecting sets of lines at oblique angles. There also may be parting parallel to $\{02\bar{2}1\}$.

Relief varies with the direction. It is high when the long diagonal of the rhomb is parallel to the vibration plane of the lower nicol and low when the short diagonal is in this position. An occasional section parallel to $\{0001\}$ has high relief in all positions.

Birefringence extreme, $n_\omega - n_\epsilon = 0.180 \text{ to } 0.190$; interference colors are pearl gray or white of the high order. Rather bright colors of the fourth and fifth orders may show on the edge of the slide.

Extinction symmetrical to outlines of crystals and to the cleavage traces. Curved crystals show wavy extinction.

Twinning.—The dolomite of metamorphic rocks usually shows polysynthetic twinning with $\{02\bar{2}1\}$ as twin-plane. The twinning lamellae are parallel to both short and long diagonals of the

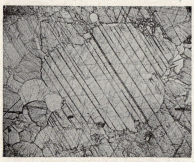

rhombs. As in calcite, the twin lamellae are usually so thin that they show first-order interference colors.

Interference Figure.—The interference figure is uniaxial negative with many rings.

Distinguishing Features.—Dolomite closely resembles calcite, but in many cases it may be distinguished by its tendency to euhedral crystals, by zonal structure, and by

Fig. 172.—($\times$10) Anhedral dolomite (in metamorphic dolomite) showing twin lamellae and cleavage.

twinning lamellae parallel to the short diagonal. It is even more like magnesite, and thus it may be necessary to rely on chemical or microchemical tests.

Occurrence.—Dolomite is a very common mineral. It occurs in veins and replacement deposits, in sedimentary dolomite rocks and limestones, and in metamorphic dolomite rocks.

Magnesite

$MgCO_3$ Hexagonal
(Rhombohedral Subsystem)

$$n_\epsilon = 1.509 \text{ to } 1.527$$
$$n_\omega = 1.700 \text{ to } 1.726$$
$$\text{Opt. } (-)$$

Color.—Colorless.

Form.—Magnesite usually occurs in anhedral to subhedral crystal aggregates. The porcelainlike microcrystalline variety has a grain size of the order of 1μ. Euhedral crystals are exceedingly rare.

Cleavage perfect rhombohedral $\{10\bar{1}1\}$ as in calcite, dolomite, and siderite except in the microcrystalline variety.

Relief changes on rotation like calcite and dolomite. It has high relief when the long diagonal of the rhomb is parallel to the vibration plane of the lower nicol and low relief when the short diagonal is in this position. An occasional section parallel to {0001} has high relief in all positions.

Birefringence extreme, $n_\omega - n_\epsilon = 0.191$ to 0.199; interference colors are pearl gray (white of the high order).

Extinction symmetrical with respect to cleavage traces.

Twinning absent as far as known.

Interference Figure.—The interference figure is uniaxial negative with many rings.

Distinguishing Features.— Magnesite is very similar to dolomite and calcite and has no distinctive optical properties of its own aside from

Fig. 173.—(×25) Subhedral magnesite crystals in metamorphic magnesite rock.

indices of refraction. For this reason chemical or microchemical tests may be necessary to distinguish it.

Occurrence.—Metamorphic magnesite rocks are found in Stevens County, Washington. Magnesite is a common mineral in serpentine as both coarsely crystalline and microcrystalline varieties.

Siderite

$FeCO_3$

Hexagonal
(Rhombohedral Subsystem)

$$n_\epsilon = 1.596 \text{ to } 1.633$$
$$n_\omega = 1.830 \text{ to } 1.875$$
$$\text{Opt. } (-)$$

Color.—In thin sections it is colorless to gray and may be yellowish or brown in spots on the edges. The brown spots are due to alteration.

Form.—Siderite occurs in fine to coarse aggregates of anhedral to subhedral crystals and sometimes shows oolitic, spherulitic, or colloform structure.

Cleavage perfect rhombohedral {$10\bar{1}1$} as in calcite, dolomite, and magnesite.

Relief varies somewhat on rotation. The relief is high when the long diagonal is parallel to the vibration plane of the lower nicol and moderate when the short diagonal is in this position. In both positions the index of refraction is greater than balsam.

Birefringence extreme, $n_\omega - n_\epsilon = 0.234$ to 0.242. Interference colors are pearl gray (white of high order). Brighter colors may show on the edge of the slide.

Extinction symmetrical to cleavage traces.

Twinning.—Twin lamellae parallel to the long diagonal [twin-plane = $\{01\bar{1}2\}$] are occasionally observed.

FIG. 174.—($\times$12) Siderite with secondary limonite (dark stain).

Interference Figure.—The interference figure is uniaxial negative with numerous rings.

Distinguishing Features.—Siderite very much resembles the other rhombohedral carbonates but may often be distinguished by the brown stain around the borders of the grains and along cleavage cracks. The index of refraction in all positions is greater than that of balsam; in calcite, dolomite, and magnesite the index of refraction n_ϵ is less than that of balsam.

Occurrence.—The chief occurrence of siderite is in veins or replacement deposits with quartz as a common associate. Siderite also is a prominent mineral in some bauxites. It is the principal mineral of septarian clay iron-stone concretions. Siderite is a prominent mineral in the oolitic iron-stones of England as an associate of chamosite. It is a secondary mineral in the cavities of some basalts.

Aragonite

CaCO₃   Orthorhombic

$$n_\alpha = 1.530$$
$$n_\beta = 1.682$$
$$n_\gamma = 1.686$$
$$2V = 18°; \text{ Opt. } (-)$$
$$a = \beta \text{ or } Y, b = \gamma \text{ or } Z, c = \alpha \text{ or } X$$

Color.—Colorless in thin sections.

Form.—Aragonite usually shows a columnar or fibrous structure. Cross sections are six sided.

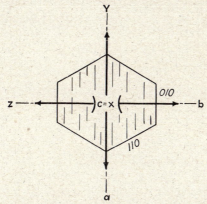

Fig. 175.—Orientation diagram of aragonite. Section parallel to (001).

Cleavage imperfect parallel to the length of the crystals (010 face).

Relief varies with the direction; the relief is low when the columns or fibers are parallel to the vibration plane of the lower nicol and high when these are normal to this direction. Basal sections show no change of relief since n_β is about the same as n_γ.

Birefringence extreme, $n_\gamma - n_\alpha = 0.156$. Interference colors pearl gray (white of the high order); brighter colors may show on thin edges and along cracks.

Extinction parallel to crystals or columns.

Twinning fairly common [twin-plane = {110}] both as twin lamellae and as contact and penetration twins.

Interference Figure.—Basal {001} sections of aragonite give a negative biaxial interference figure with a small axial angle. The axial plane is {100}. Dispersion, $r < v$ weak.

Distinguishing Features.—Aragonite greatly resembles calcite but lacks the rhombohedral cleavage. It is biaxial, whereas calcite is uniaxial.

FIG. 176.—(×45) Aragonite in palagonite tuff. (Note difference in relief of various crystal sections.)

Alteration.—Aragonite alters easily to calcite, which is the stable form of calcium carbonate.

Occurrence.—The most common occurrence of aragonite is probably as a secondary mineral in cavities of basalts and andesites. It also occurs in seams of limestones, sandstones, and occasionally in veinlets. It was probably a widespread original constituent of sediments but has since been altered to calcite. It is also found in some metamorphic rocks.

Barite

BaSO₄   Orthorhombic

$$n_\alpha = 1.636$$
$$n_\beta = 1.637$$
$$n_\gamma = 1.648$$
$$2V = 36 \text{ to } 37\tfrac{1}{2}°; \text{ Opt. } (+)$$
$$a = \gamma \text{ or } Z, \ b = \beta \text{ or } Y, \ c = \alpha \text{ or } X$$

Color.—Colorless in thin sections.

Form.—Usually in granular aggregates, but the individual crystals may be elongate.

Cleavage in three directions, parallel to {001} and {110} and therefore at angles of 90 and 78°.

Relief fairly high, $n >$ balsam.

Birefringence rather weak, $n_\gamma - n_\alpha = 0.012$, slightly greater than that of quartz. The maximum interference color is rarely above first-order yellow or orange. The interference colors are frequently mottled.

Extinction parallel to the best cleavage {001}. The extinction in {001} sections is symmetrical.

Orientation.—The direction of the best cleavage is the slower ray.

Twinning.—Polysynthetic twinning with {110} as the twin-plane is occasionally found.

Interference Figure.—Sections cut parallel to {100} give a positive biaxial interference figure with a moderate axial angle. Cleavage plates parallel to {001} give an obtuse bisectrix figure. The axial plane is {010}. Dispersion, $r < v$ weak.

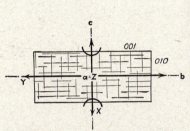

Fig. 177.—Orientation diagram of barite. Section parallel to (100).

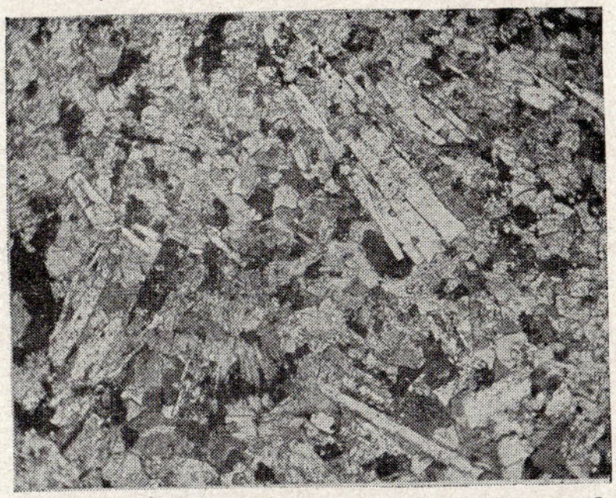

Fig. 178.—(×30) Barite (bladed) with calcite in limestone.

Distinguishing Features.—Barite greatly resembles celestite, but the axial angle is smaller. It may be necessary to determine refractive indices carefully or to make chemical tests in order to distinguish them.

Occurrence.—Barite is a prominent vein mineral; the common associates are quartz and calcite. It also occurs in limestones and sandstones and is prominent in some concretions, but it is rare as a strictly rock-forming mineral.

Celestite

$SrSO_4$ Orthorhombic

$$n_\alpha = 1.622$$
$$n_\beta = 1.624$$
$$n_\gamma = 1.631$$
$$2V = 51°; \text{ Opt. } (+)$$
$$a = \gamma \text{ or } Z, b = \beta \text{ or } Y, c = \alpha \text{ or } X$$

Color.—Colorless in thin sections.

Form.—Euhedral to anhedral crystals, sometimes fine granular. Euhedral crystals are mostly tabular parallel to {001} and elongated in the direction of the b-axis [010].

Cleavage perfect parallel to {001}, imperfect parallel to {110}.

Relief fair, $n >$ balsam.

Birefringence rather weak, $n_\gamma - n_\alpha = 0.009$, the same as that of quartz, so that the highest interference color is white or straw yellow.

Extinction parallel to the outlines and to the cleavage.

Orientation.—The elongation of tabular crystals is parallel to the slower ray.

Interference Figure.—Sections cut parallel to {100} give a positive biaxial interference figure with moderate axial angle.

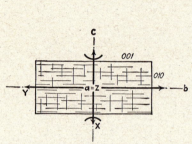

Fig. 179.—Orientation diagram of celestite. Section parallel to (100).

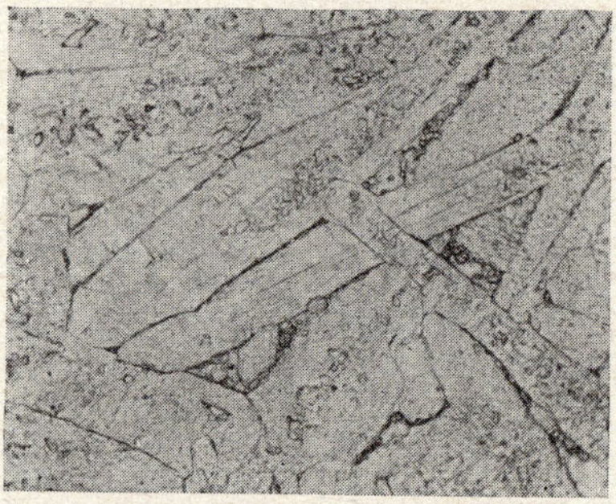

Fig. 180.—(×30) Celestite. (Random section of a crystal aggregate.)

Cleavage plates parallel to {001} give an obtuse bisectrix figure. The axial plane is {010}. Dispersion, $r < v$.

Distinguishing Features.—Celestite very much resembles barite, but the axial angle is larger.

Occurrence.—Celestite usually occurs in sedimentary limestones, where it is probably more common than barite.

Anhydrite

CaSO₄   Orthorhombic

$$n_\alpha = 1.570$$
$$n_\beta = 1.576$$
$$n_\gamma = 1.614$$
$$2V = 42°; \text{ Opt. } (+)$$
$$a = \gamma \text{ or } Z,\ b = \beta \text{ or } Y,\ c = \alpha \text{ or } X$$

Color.—Colorless in thin sections.

Form.—Usually fine to medium-grained aggregates or anhedral to subhedral crystals, which are sometimes elongate. Euhedral crystals are rare. It also occurs as inclusions in halite.

Cleavage in three directions at right angles parallel to {100}, {010}, and {001}. It may also show parting parallel to {101} which is due to twinning.

Relief moderate, $n >$ balsam. Some sections show a slight change of relief when the stage is rotated.

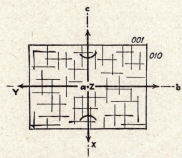

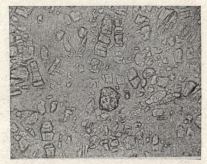

Fig. 181.—Orientation diagram of anhydrite. Section parallel to (100).

Fig. 182.—(×20) Anhydrite partially altered to gypsum. The crystal in the center with high relief is dolomite.

Birefringence strong, $n_\gamma - n_\alpha = 0.044$. The interference colors range up to about third-order green.

Extinction parallel to the cleavage traces.

Twinning.—Polysynthetic twinning with {101} as twin-plane is common. The twinning lamellae show best on the (010) face and make an angle of 42 and 48° with the cleavage traces. There may be two sets of twin lamellae (101) and ($\bar{1}$01) intersecting at angles of $83\frac{1}{2}$ and $96\frac{1}{2}$°. (See Fig. 183.)

Interference Figure.— Cleavage fragments and sections parallel to {100} give a biaxial positive interference figure with a moderate axial angle. The axial plane is {010}. Dispersion, $r < v$.

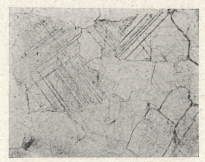

Fig. 183.—(×30) Anhydrite showing cleavage and twin lamellae in metamorphic anhydrite rock.

Distinguishing Features.—Anhydrite is distinguished from gypsum by higher relief and stronger birefringence. The rectangular pseudo-cubic cleavage is distinctive.

Alteration.—Between grains and along veinlets anhydrite is often altered to gypsum, and anhydrite may be found as remnants within gypsum.

Occurrence.—Anhydrite occurs in sedimentary beds. It is often encountered in deep drilling but near the surface is usually altered to gypsum. It often occurs with halite and is common in salt mines. In salt domes it is often found as cap-rock. Metamorphic anhydrite rock is found at the Nevada-Douglas mine, Lyon County, Nevada. It is occasionally encountered in veins formed at depth.

GYPSUM

$CaSO_4.2H_2O$

Monoclinic
$\angle \beta = 80°42'$

$$n_\alpha = 1.520$$
$$n_\beta = 1.522$$
$$n_\gamma = 1.529$$
$$2V = 58°; \text{ Opt. } (+)$$
$$b = \beta \text{ or } Y, c \wedge \alpha \text{ or } X = +37°28'$$

Color.—Colorless in thin sections.

Form.—Gypsum usually occurs in anhedral to subhedral aggregates and is often uneven grained. It sometimes shows a fibrous structure.

Cleavage perfect in one direction {010}, imperfect parallel to {100} and {$\bar{1}11$}. ($\bar{1}00$) $\wedge$ ($\bar{1}01$) = 66°10'.

Relief low, n slightly < balsam.

Birefringence rather weak, $n_\gamma - n_\alpha = 0.009$ (the same as that of quartz). The highest interference color is white or straw yellow. Sections with the highest interference color do not usually show any cleavage.

Extinction parallel to the best cleavage in sections parallel to (010).

Orientation.—Cleavage traces are parallel to both slower and faster rays since $b = \beta$ or Y.

Twinning.—The polysynthetic twinning often found in thin sections of gypsum is produced by heating the section.

Interference Figure.—The interference figure is biaxial positive with a moderate axial angle. The axial plane is {010}. Dispersion, $r > v$. Sections parallel to (010) give a "flash figure."

Distinguishing Features. Gypsum is easily distinguished from anhydrite by lower relief and weaker birefringence.

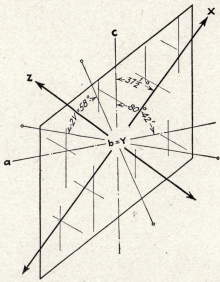

Fig. 184.—Orientation diagram of gypsum. Section parallel to (010).

Occurrence.—Gypsum is the chief constituent of gypsum rock, which in most cases has been formed by the hydration of anhydrite. Anhydrite may occur in the gypsum as a relict

Fig. 185.—(×12) Gypsum in gypsum rock. (× nicols.)

mineral. Gypsum occurs in veinlets and between grains of anhydrite. Other commonly associated minerals are calcite, dolomite, and halite.

Polyhalite

$K_2MgCa_2(SO_4)_4.2H_2O$ Triclinic

$$\angle\alpha = 92°29'$$
$$\angle\beta = 123°4'$$
$$\angle\gamma = 88°21'$$

$$n_\alpha = 1.548$$
$$n_\beta = 1.562$$
$$n_\gamma = 1.567$$
$$2V = ca.\ 70°;\ Opt.\ (-)$$

(Optical orientation unknown)

Color.—Colorless to reddish in thin sections. The reddish color is due to hematitic pigment.

Form.—Polyhalite shows granular or fibrous structure.

Cleavage parallel to (100) and parting parallel to (010).

Relief low, $n >$ balsam.

Birefringence moderate, $n_\gamma - n_\alpha = 0.019$. The interference colors range up to second-order blue.

Extinction oblique.

Twinning.—Polysynthetic twinning with (010) as the twin-plane is very common.

Interference Figure.—The interference figure is biaxial negative with a rather large axial angle, but it may be difficult to obtain on account of the small size of the crystals.

FIG. 186.—(×9) Polyhalite showing polysynthetic twinning and variation in pigment. (× nicols.)

Distinguishing Features.—Polyhalite may resemble gypsum, but both its refractive indices and birefringence are higher. It is decomposed by water with the separation of microchemical gypsum.

Occurrence.—Polyhalite occurs in saline beds; the common associates are halite, sylvite, magnesite, and anhydrite. In the West Texas–New Mexico Permian basin it is an alteration product of anhydrite according to Schaller and Henderson.

Alunite

$KAl_3(OH)_6(SO_4)_2$ Hexagonal
                                                    (Rhombohedral Subsystem)

$$n_\omega = 1.572$$
$$n_\epsilon = 1.592$$
$$\text{Opt. } (+)$$

Color.—Colorless in thin sections.

Form.—Alunite usually shows fine to coarse aggregates. Crystals vary from tabular to pseudo-cubic rhombohedral ($rr' = 90°50'$).

Cleavage.—Fair cleavage in one direction {0001}.

Relief f a i r, $n >$ balsam. When the stage is rotated there is a slight change of relief.

Birefringence m o d e r a t e, $n_\epsilon - n_\omega = 0.020$; the interference colors range up to second-order blue.

Extinction parallel or symmetrical in most sections. Basal sections are dark in all positions.

Fig. 187.—(×15) Alunite showing aggregate structure. (× nicols.)

Orientation.—Crystals and cleavage traces are length-fast.

Interference Figure.—Basal sections give a positive uniaxial interference figure.

Related Minerals.—Natroalunite, the sodium analogue of alunite, is very similar to alunite in its properties.

Occurrence.—Alunite occurs as a hydrothermal alteration product of rhyolites, dacites, and andesites. It is also prominent as a vein mineral.

Jarosite

$KFe_3^{III}(OH)_6(SO_4)_2$

$$n_\epsilon = 1.715$$
$$n_\omega = 1.820$$
$$\text{Opt. } (-)$$

Color.—Colorless to brown in thin sections.

Form.—Jarosite occurs in crystal aggregates and occasionally in euhedral crystals, which are similar to those of alunite, for these two minerals are isomorphous.

Cleavage.—Distinct cleavage in one direction {0001}.

Relief very high, $n >$ balsam.

Birefringence extreme, $n_\omega - n_\epsilon = 0.105$.

Extinction parallel or symmetrical. Basal sections are dark in all positions.

Orientation difficult to test on account of the extreme birefringence.

Interference Figure.—Tabular crystals give a negative uniaxial figure with many rings.

Related Minerals.—Natrojarosite, $NaFe_3(OH)_6(SO_4)_2$; ammoniojarosite $NH_4Fe_3(OH)_6(SO_4)_2$; plumbojarosite $PbFe_6(OH)_{12}(SO_4)_4$; and argentojarosite $AgFe_3(OH)_6(SO_4)_2$ are all very similar to jarosite.

Alteration.—Jarosite alters readily to limonite.

Occurrence.—Jarosite is a rather common mineral in the lower oxidized zone of ore deposits. It is occasionally found in volcanic igneous rocks, perhaps as a late hydrothermal mineral.

Monazite

$(Ce,La,Nd,Pr)PO_4$

Monoclinic
$\angle\beta = 76°6'$

$$n_\alpha = 1.786 \text{ to } 1.800$$
$$n_\beta = 1.788 \text{ to } 1.801$$
$$n_\gamma = 1.837 \text{ to } 1.849$$
$$2V = 6 \text{ to } 19°; \text{ Opt. } (+)$$
$$b = \alpha \text{ or } X, \ c \wedge \gamma \text{ or } Z = -2 \text{ to } -10°$$

Color.—Nearly colorless to neutral in thin sections.

Form.—Monazite occurs in euhedral crystals, which are usually very small.

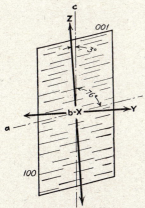

Fig. 188.—Orientation diagram of monazite. Section parallel to (010).

Fig. 189.—(×10) Monazite crystals in quartz matrix.

Cleavage.—Parting parallel to {001} is often prominent.

Relief very high, $n >$ balsam.

Birefringence strong to very strong, $n_\gamma - n_\alpha = 0.049$ to 0.051. The maximum interference color is upper third or lower fourth order. Cross sections of crystals have very weak birefringence since $n_\beta - n_\alpha = 0.001$ to 0.002.

Extinction.—Longitudinal sections have a small extinction angle (2 to 10°). Sections parallel to {001} do not show complete extinction.

Orientation.—Crystals are length-slow.

Interference Figure.—The interference figure is biaxial positive with a small axial angle. The axial plane is normal to {010}. Dispersion strong, $r < v$.

Distinguishing Features.—Monazite is more like sphene than any common mineral, but its birefringence is not so high.

Occurrence.—Monazite occurs in some granite pegmatites. It is a characteristic detrital mineral but is rare except in certain sands where there has been a concentration of the mineral.

APATITE

$3Ca_3(PO_4)_2.CaF_2$

Hexagonal
(Hexagonal Subsystem)

$$n_\epsilon = 1.630 \text{ to } 1.651$$
$$n_\omega = 1.633 \text{ to } 1.655$$
$$\text{Opt. } (-)$$

Color.—Colorless in thin sections.

Form.—Apatite is usually found as minute six-sided prismatic crystals. It is a common and widely distributed mineral but usually occurs in small amounts.

Cleavage imperfect basal {0001} shown as cross fractures.

Fig. 190.—(×20) Apatite crystals in an igneous rock.

Larger crystals may show imperfect cleavage parallel to the length {10$\bar{1}$0}.

Relief moderate, $n >$ balsam.

Birefringence weak, $n_\omega - n_\epsilon = 0.003$ to 0.004. The interference colors are first-order gray to white. Cross sections are dark between crossed nicols.

Extinction parallel.

Orientation.—The crystals are usually length-fast, but crystals of tabular habit are length-slow.

Interference Figure.—Basal sections are usually too small to give good interference figures.

Distinguishing Features.—Apatite is distinctive. The only common mineral that closely resembles it is dahllite, which occurs as a secondary mineral in cavities and seams associated with collophane.

Related Minerals.—Wilkeite, a rare mineral of the apatite group with the sulfate radical; voelckerite or oxy-apatite; fermorite, strontian apatite, and ellestadite are similar to apatite in physical properties.

Occurrence.—Apatite is a common minor accessory mineral of practically all igneous rocks. In the opinion of Tolman and Rogers, it is a late magmatic mineral and not an early one. It also occurs in pegmatites, in some high-temperature veins, and in metamorphic limestones and is also prominent in some iron ores.

Dahllite

$3Ca_3(PO_4)_2.CaCO_3$ Hexagonal
(Hexagonal Subsystem)

$$n_\epsilon = 1.619 \text{ to } 1.626$$
$$n_\omega = 1.623 \text{ to } 1.635$$
$$\text{Opt. } (-)$$

Color.—Colorless to pale brown or gray in thin sections.

Form.—Dahllite occurs in minute hexagonal crystals, in crusts with banded subradiating structure, in spherulites, and in fine-grained aggregates forming concretions or sedimentary rocks.

Relief moderate, $n >$ balsam.

Birefringence weak, $n_\omega - n_\epsilon = 0.004$ to 0.009. Interference colors are bluish gray to white of the first order.

Extinction parallel. Cross sections are dark between crossed nicols, but occasionally they may show biaxial sectors.

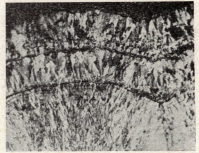

FIG. 191.—($\times$10) Dahllite incrustation. ($\times$ nicols.)

Orientation.—Prismatic crystals are length-fast like apatite. The columns of crusts and fibers of spherulites are also length-fast. Sections of tabular crystals are length-slow.

Interference Figure.—Basal sections are usually too small to give good interference figures.

Distinguishing Features.—Dahllite is much like apatite, but it is always a secondary mineral. To make certain of its identity it may be necessary to try the solubility of isolated particles of the mineral.

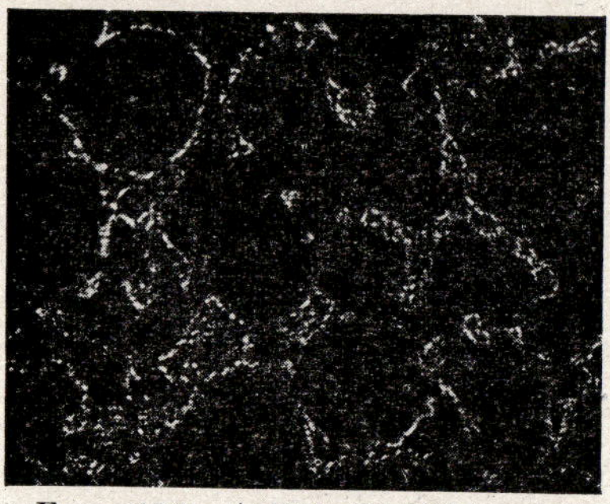

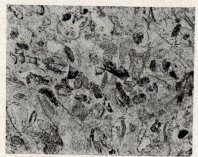

Related Minerals.—Francolite is a closely related mineral.

Occurrence.—Dahllite occurs as a secondary mineral in phosphorite or so-called phosphate rock. The usual associate is collophane. The dahllite has probably been formed by the gradual crystallization of the collophane and by the migration of some of the calcium phosphate.

FIG. 192.—(×30) Dahllite rims around collophane in phosphorite. (× nicols.)

COLLOPHANE

$$3Ca_3(PO_4)_2 . nCa(CO_3, F_2, O)(H_2O)_x$$

Amorphous
(Mineraloid)

$$n = 1.57 \text{ to } 1.62$$

Amorphous calcium carbonophosphate is usually considered to be a massive form of apatite, but it is distinctive and should be listed separately as a mineraloid.

Color.—Usually light to dark brown, yellowish brown, gray, etc., in thin sections, but occasionally it is colorless.

Form.—Collophane is usually massive but may be oolitic or colloform, in grains and fragments. It often shows the organic structure of bones, molluscs, brachiopods, crinoids, bryozoans, or corals.

FIG. 193.—(×17) Collophane (dark) with calcite (light) in phosphatic limestone.

Cleavage absent. Irregular fracture may show on edges of the slide.

Relief moderate, $n >$ balsam. The index of refraction is variable, but it is usually 1.60 to 1.61.

Birefringence.—Usually isotropic but may show weak form-birefringence (up to 0.005). Pseudo-spherulitic structure (concentric instead of fibrous elements) sometimes shows.

Orientation.—Birefringent areas may be length-slow or length-fast.

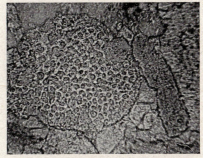

FIG. 194.—($\times$117) Collophane (gray) and calcite (white) in phosphatic limestone showing bryozoan on left and ostracoderm on right.

Distinguishing Features.—Some specimens of collophane resemble opal, but the refractive index of the latter is always less than that of balsam. Oolitic chamosite resembles oolitic collophane.

Alteration.—Collophane is often more or less replaced by calcite. Replacement by quartz, chalcedony, or opal is very rare. In some specimens dahllite seems to be forming at the expense of collophane.

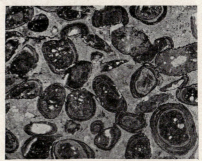

FIG. 195.—($\times$12) Collophane in oolitic phosphorite. (Both the matrix and the ooliths are collophane.)

Occurrence.—In sedimentary phosphatic limestones, in phosphorites or so-called phosphate rocks as the chief constituent, and in phosphate nodules. It is the dominant mineral of fossil bone, in which the microstructure of the original bone is usually preserved (see Fig. 196). In fossil bone it has been formed by phosphatic enrichment. In invertebrate fossils it has been formed by enrichment in the case of

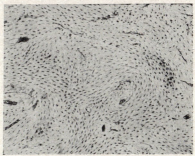

FIG. 196.—($\times$12) Collophane (fossil bone) showing Haversian canals and lacunae of the original bone.

phosphatic brachiopods such as *Lingula*, but in most cases by the replacement of original calcareous organisms.

It also occurs as a detrital mineral in beach sands of the South Atlantic states according to J. H. C. Martens.

Lazulite

$Al_2(Mg,Fe)(OH)_2(PO_4)_2$ Monoclinic
$\angle \beta = 88°$

$$n_\alpha = 1.603 \text{ to } 1.604$$
$$n_\beta = 1.632 \text{ to } 1.633$$
$$n_\gamma = 1.639 \text{ to } 1.642$$
$$2V = ca. \; 69°; \text{ Opt. } (-)$$
$$b = \beta \text{ or } Y, \; c \wedge \alpha \text{ or } X = +9°$$

Color.—Blue to colorless in thin sections. Some sections are pleochroic from blue to colorless. Axial colors: α or X = colorless; β or Y = azure blue; γ or Z = azure blue.

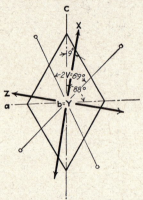

FIG. 197.—Orientation diagram of lazulite. Section parallel to (010).

FIG. 198.—($\times 12$) Lazulite showing polysynthetic twinning with included patches of quartz. ($\times$ nicols.)

Form.—Lazulite is occasionally found in euhedral crystals of bipyramidal habit but it usually occurs in anhedra.

Cleavage indistinct parallel to {110}.

Relief fairly high, $n >$ balsam.

Birefringence strong, $n_\gamma - n_\alpha = 0.036$ to 0.038, so that the maximum interference color is upper second or lower third order.

Extinction oblique.

Orientation.—The long diagonal of the crystal sections is the faster ray.

Twinning.—Polysynthetic twinning is common. Twin-axis = [001].

Interference Figure.—The figure is biaxial negative with a large axial angle. The axial plane is {010}. Dispersion, $r < v$.

Distinguishing Features.—Lazulite is practically the only blue pleochroic mineral with strong birefringence.

Occurrence.—As far as known, lazulite is confined to metamorphic rocks. It occurs in quartzites and in quartz veins. The usual associates are quartz, rutile, corundum, pyrophyllite, kyanite, and andalusite.

Perovskite

$CaTiO_3$ Pseudo-isometric

$$n = 2.34 \text{ to } 2.38$$

Color.—Yellow to brown in thin sections.

Form.—Perovskite is usually found in minute cubic crystals.

Cleavage cubic, noticed only in the larger crystals.

Relief very high, $n >$ balsam. It is difficult to make the Becke test on account of total reflection. In reflected light it shows adamantine luster.

Birefringence nil to 0.002. Minute crystals are dark between crossed nicols; larger crystals show very weak birefringence.

Twinning.—The larger crystals show complicated polysynthetic twinning.

Distinguishing Features.—Perovskite resembles melanite (garnet) and picotite (spinel), but it has a much higher refractive index than either of these.

Occurrence.—Perovskite is a rare, but widely distributed, mineral in basic igneous rocks, especially melilite basalts and peridotites. It is also found in chlorite and talc schists and in some metamorphic limestones.

THE FELDSPARS

Of all the silicates the feldspars are the most important rock-forming minerals since they constitute about 60 per cent of the earth's crust or outer shell. The classification of igneous rocks depends to a large extent on the character of the feldspar.

There are three well-marked groups of feldspars. The first group consists of the monoclinic potassium feldspar, orthoclase, and its dimorph, sanidine. The rare barium feldspar, celsian, $BaAl_2Si_2O_8$, also is monoclinic.

The microcline group is triclinic with microcline and the closely related anorthoclase, sometimes called *soda microcline*.

Next we have the triclinic group of plagioclases, which are isomorphous mixtures of two end members, albite and anorthite. The plagioclase group will be discussed in detail later on.

THE FELDSPARS

Mineral	Chemical composition	n_α	n_β	n_γ
Orthoclase	$(K,Na)AlSi_3O_8$	1.519	1.524	1.526
Sanidine	$(K,Na)AlSi_3O_8$	1.517–1.520	1.523–1.525	1.524–1.526
Microcline	$KAlSi_3O_8$	1.518–1.522	1.522–1.526	1.525–1.530
Anorthoclase	$(Na,K)AlSi_3O_8$	1.522–1.536	1.526–1.549	1.527–1.549
Plagioclase { Albite	$NaAlSi_3O_8$	1.525–1.532	1.529–1.536	1.536–1.541
Anorthite	$CaAl_2Si_2O_8$	1.571–1.575	1.577–1.583	1.582–1.588

The feldspars are monoclinic or triclinic pseudo-monoclinic. There is perfect cleavage in two directions {001} and {010} at 90 or nearly 90° (about 86 to 87°). Cleavage fragments appear as in Figs. 199 *a,b*. The feldspars have low relief in balsam and birefringence varying from 0.005 to 0.013. They are all biaxial, with a very small axial angle for sanidine, a moderate one for anorthoclase, and a large to very large one for orthoclase, microcline, and the plagioclases.

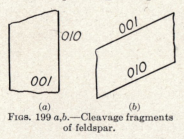

(a)   (b)
FIGS. 199 *a,b*.—Cleavage fragments of feldspar.

Twinning is common in all the feldspars and is almost universal in microcline and the plagioclases. The extinction angles in twinned crystals constitute one of the chief methods of distinguishing the individual members of these groups.

ORTHOCLASE

(K,Na)AlSi₃O₈

Monoclinic

$\angle\beta = 63°57'$

$$n_\alpha = 1.518$$
$$n_\beta = 1.524$$
$$n_\gamma = 1.526$$
$$2V = 69 \text{ to } 72°; \text{ Opt. } (-)$$
$$b = \gamma \text{ or } Z, a \wedge \alpha \text{ or } X = +5 \text{ to } +12°,$$
$$c \wedge \beta \text{ or } Y = -14 \text{ to } -21°$$

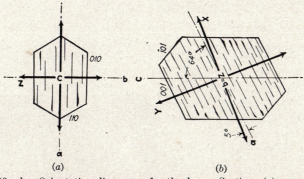

(a)   (b)

Figs. 200 a,b.—Orientation diagrams of orthoclase. Sections (a) normal to the c-axis and (b) parallel to (010).

Color.—Colorless in thin sections, but may be cloudy on account of incipient alteration in contrast with quartz, which is clear.

Form.—Orthoclase occurs in phenocrysts, in subhedral and anhedral crystals, and in spherulites.

Cleavage.—Perfect cleavage parallel to {001}, less perfect parallel to {010}, imperfect parallel to {110}.

Fig. 201.—(×37) Orthoclase. Carlsbad twins. (× nicols.)

Relief low, $n <$ balsam.

Birefringence weak, $n_\gamma - n_\alpha = 0.008$; so the interference colors are gray and white of the first order and the maximum a little lower than that of quartz in the same slide.

Extinction on {001} parallel, on {010} from 5 to 12°, increasing with the soda content.

Orientation.—Cleavage traces on {010} make a small angle with the faster ray.

Fig. 202.—(×80) Orthoclase-cristobalite spherulite in volcanic glass.

Twinning.—Twinning according to the Carlsbad law (*c*-axis or [001] = twin-axis). These are simple twins consisting of two individuals and are practically never polysynthetic.

Interference Figure.—The interference figure is biaxial negative with a large axial angle. The axial plane is normal to {010}. Dispersion, $r > v$.

Distinguishing Features.—Orthoclase is distinguished from its dimorph sanidine by its large axial angle. Sanidine practically always occurs in phenocrysts.

Occurrence.—Orthoclase is a widely distributed mineral in persilicic igneous rocks such as granites and syenites. In spherulites of obsidian and rhyolite it is often intergrown with cristobalite or quartz. The potassium feldspar of metamorphic rocks is usually microcline.

Fig. 203.—(×9) Orthoclase overgrowth on plagioclase in trachyte porphyry. (× nicols.)

It is also common in detrital deposits and in sandstones and arkoses.

Adularia

KAlSi₃O₈   Monoclinic

Although adularia is probably a variety of orthoclase and not
a distinctive mineral, it may well be treated separately.

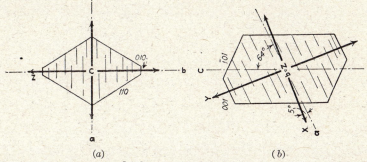

(a)	(b)

FIGS. 204 *a,b.*—Orientation diagrams of adularia. Sections (*a*) normal to the
c-axis and (*b*) parallel to (010).

The optical properties of adularia are the same as those of
orthoclase, but there is a distinction in crystal habit.   Adularia

FIG. 205.—(×31) Adularia in vein
quartz.

FIG. 206.—(×56) Adularia overgrowth on orthoclase along quartz-
adularia—fluorite veinlet in altered
phonolite (opaque mineral is pyrite).

is pseudo-orthorhombic with a rhombic cross section (110 ∧ 1̄10
= 61°13′).   The (010) face is very narrow or absent.

Adularia is a rather low-temperature feldspar found in veins
and replacement deposits, and in some rocks of low-grade metamorphism.   It is especially characteristic of Tertiary gold and
silver ores of the bonanza type.   The crystals are commonly
minute and can be identified only with a rather high-power
objective (8 mm).

SANIDINE

$(K,Na)AlSi_3O_8$

Monoclinic
$\angle\beta = 63°57'$

$$n_\alpha = 1.517 \text{ to } 1.520$$
$$n_\beta = 1.523 \text{ to } 1.525$$
$$n_\gamma = 1.524 \text{ to } 1.526$$
$$2V = 0 \text{ to } 12°; \text{ Opt. } (-)$$

Orientation: (1) Ax. pl. {010}, $b = \beta$ or Y, $a \wedge \alpha$ or X $= +5°$ or (2) Ax. pl. $\perp$ {010}, $b = \gamma$ or Z, $a \wedge \alpha$ or X $= +5°$

Color.—Colorless in thin sections; clear in contrast with orthoclase, which is often cloudy.

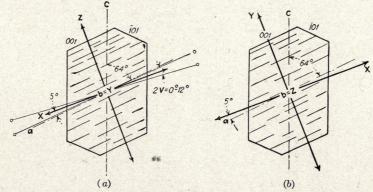

Figs. 207 *a,b.*—Orientation diagrams of sanidine. Sections parallel to (010).

Form.—Sanidine usually occurs in distinct euhedral crystals as phenocrysts.

Cleavage perfect parallel to {001}, less perfect parallel to {010}. There may also be parting parallel to {100}.

Relief low, $n <$ balsam.

Birefringence weak, $n_\gamma - n_\alpha = 0.007$, so the interference colors are gray and grayish white of the first order.

Extinction on (001) parallel, on (010) $+5°$. Sections normal to an optic axis remain practically dark since the axial angle is often very small.

Twinning.—Usually according to the Carlsbad law (*c*-axis or [001] = twin-axis). Twins are simple twins of two individuals and are rarely polysynthetic.

Interference Figure.—Some sections give a negative biaxial interference figure with a small axial angle, but the angle may be so small that the figure is almost uniaxial. Dispersion, (1) $r < v$, (2) $r > v$.

Distinguishing Features.—Sanidine is distinguished from orthoclase by the small axial angle and in some cases by a difference of orientation. Orthoclase is usually cloudy on account of incipient alteration; sanidine, on the other hand, is clear.

Occurrence.—Sanidine is practically confined to persilicic volcanic rocks such as rhyolites and trachytes and the corresponding tuffs.

MICROCLINE

$KAlSi_3O_8$ Triclinic

$$\angle\alpha = 89°53'$$
$$\angle\beta = 64°10'$$
$$\angle\gamma = 90°51'$$

$$n_\alpha = 1.518 \text{ to } 1.522$$
$$n_\beta = 1.522 \text{ to } 1.526$$
$$n_\gamma = 1.525 \text{ to } 1.530$$
$$2V = 77 \text{ to } 84°; \text{ Opt. } (-).$$

Ax. pl. and γ or Z are nearly $\perp (010)$. Angle between trace of ax. pl. and edge (001): (010) $= +5°$

Color.—Colorless in thin sections but may be cloudy on account of incipient alteration.

Form.—Microcline is usually found in subhedral to anhedral crystals. Euhedral crystals are rarely seen in rock sections.

Cleavage perfect parallel to {001}, less perfect parallel to {010}, imperfect parallel to {110} and {1$\bar{1}$0}.

Relief low, $n < $ balsam.

Birefringence weak, $n_\gamma - n_\alpha = 0.007$, so interference colors are gray and white of the first order.

Extinction.—Extinction angle on (001) $= +15°$, on (010) $= +5°$.

Orientation.—Cleavage traces on (010) are about parallel to the faster ray.

Twinning.—Polysynthetic twinning is almost universal in microcline. The twinning is in two directions, one according to the albite law ({010} = twin-plane) and the other according to the pericline law (*b*-axis or [010] = twin-axis). This usually gives the so-called *gridiron* or *quadrille* structure, the two sets of lamellae being at right angles. The twin lamellae are usually spindle shaped and the extinction usually wavy.

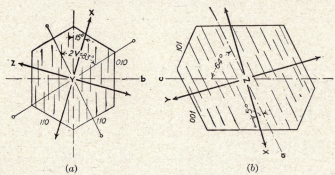

Figs. 208 *a,b*.—Orientation diagrams of microcline. Sections (*a*) normal to the *c*-axis and (*b*) parallel to (010).

Intergrowth.—Albite is commonly intergrown with microcline so that the (010) directions are parallel. This intergrowth is known as *perthite*.

Interference Figure.—On account of the twinning it is usually difficult to obtain good interference figures. Dispersion, $r > v$.

Fig. 209.—(×9) Microcline section oriented parallel to (001). There are several albite inclusions probably formed by replacement. (× nicols.)

Distinguishing Features.—Microcline is distinguished from orthoclase by the polysynthetic twinning and from both orthoclase and albite by the extinction angle of 15° on (001) and by the spindle-shaped twin lamellae.

Occurrence.—Microcline occurs in some granites, syenites, and gneisses. In the intergrowth known as *perthite* it is the principal feldspar of granite pegmatites. It is also a common mineral in sandstones, arkoses, etc., and is found as a detrital mineral in sands.

Anorthoclase

$(Na,K)AlSi_3O_8$ (Soda Microcline)   Triclinic
$$\angle\alpha = 90°6'$$
$$\angle\beta = 63°42'$$
$$\angle\gamma = 90°17'$$

$$n_\alpha = 1.522 \text{ to } 1.536$$
$$n_\beta = 1.526 \text{ to } 1.539$$
$$n_\gamma = 1.527 \text{ to } 1.541$$
$$2V = 43 \text{ to } 54°; \text{ Opt. } (-).$$
Ax. pl. nearly $\perp$ to {010}

Color.—Colorless in thin sections.

Form.—Anorthoclase occurs in phenocrysts and in anhedral crystals; also in large cleavage masses.

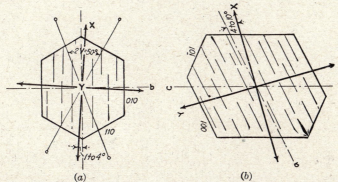

Figs. 210 *a,b.*—Orientation diagram of anorthoclase. Sections (*a*) normal to the *c*-axis and (*b*) parallel to (010).

Cleavage perfect parallel to {001}, less perfect parallel to {010}, as in the other feldspars.

Relief low, $n <$ balsam.

Birefringence weak, $n_\gamma - n_\alpha = 0.005–0.007$; the interference colors are gray and white of the first order.

Extinction on (001) = $+1$ to $+4°$, on (010) = $+4$ to $+10°$.

Twinning.—Polysynthetic twinning in two directions like that of microcline, but the lamellae are finer. It may be necessary to have a very thin section in order to detect the twinning.

Interference Figure.—The figure is biaxial negative with a moderate axial angle. Dispersion, $r > v$.

Distinguishing Features.—Anorthoclase may be distinguished from practically all other feldspars by the axial angle of about 50° (sanidine is lower and the others are higher). The small extinction angle on (001) distinguishes it from microcline and all the plagioclases except albite.

Occurrence.—The characteristic occurrence of anorthoclase is in soda-rich igneous rocks such as rhomb porphyries. It is sometimes found in pegmatites. It is a comparatively rare mineral.

PLAGIOCLASE GROUP

The minerals of this group consist of isomorphous mixtures of the two end members: albite, $NaAlSi_3O_8$, abbreviated Ab, and anorthite, $CaAl_2Si_2O_8$, abbreviated An. The following is the decimal classification of Calkins now generally used by petrographers.

Albite	Ab_{10} – Ab_9An_1
Oligoclase	Ab_9An_1 – Ab_7An_3
Andesine	Ab_7An_3 – Ab_5An_5
Labradorite	Ab_5An_5 – Ab_3An_7
Bytownite	Ab_3An_7 – Ab_1An_9
Anorthite	Ab_1An_9 – An_{10}

The plagioclases crystallize in the triclinic system with angles much like those of orthoclase. The interfacial angle (001:010) varies from 86°6′ for Ab to 85°48′ for An. The angle between the *a*-axis and *b*-axis varies from 87°5′ for Ab to 91°34′ for An, with $\gamma = 90°$ for $Ab_{58}An_{42}$.

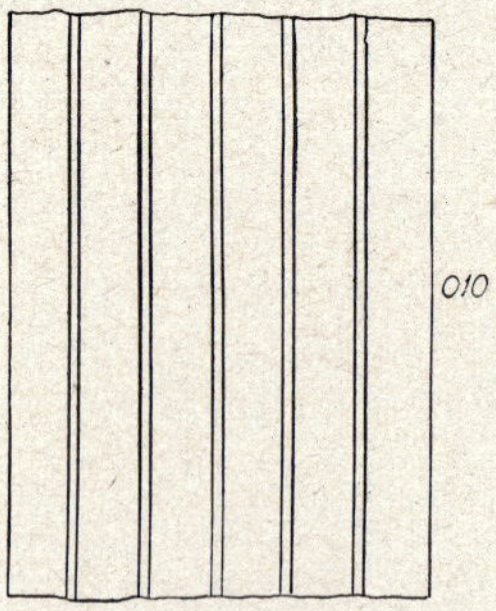

FIG. 211.—Plagioclase cleavage parallel to (001) showing albite twin-lamellae.

The cleavage of the plagioclases is perfect parallel to {001}, less perfect parallel to {010}, and imperfect parallel to {110} and {1$\bar{1}$0}.

Twinning is almost universal in the plagioclases. There are numerous twin-laws known for plagioclase, but the three most important are:

Albite law: {010} = twin-plane, (010) = composition face, Fig. 211.

Pericline law: [010] = twin-axis, rhombic section = composition face, Fig. 212.

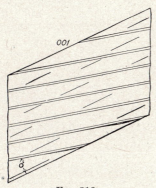

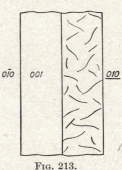

FIG. 212.

FIG. 213.

FIG. 212.—Plagiclase cleavage parallel to (010) showing pericline twin-lamellae. The angle σ is the "angle of the rhombic section."

FIG. 213.—Sketch showing Carlsbad twinning.

Carlsbad law: [001] = twin-axis, (010) = composition face, Fig. 213.

The axial angle of plagioclase varies from about 74 to 90°. The optical character is positive, negative, or neutral (2V = 90°) for a few in passing from positive to negative.

In specimens twinned according to the albite law, the lamellae appear on the (001) cleavage face and are *always* parallel to the (001 : 010) edge as shown in Fig. 211, above.

EXTINCTION ANGLES IN ORIENTED SECTIONS

Cleavage flakes of plagioclase are parallel to either (001) or (010). The former

FIG. 214.—(×20) Zonal structure in plagioclase. Section parallel to (010). (× nicols.)

are more frequent than the latter since they are parallel to the better cleavage. Albite twin lamellae are usually present on these flakes. Flakes parallel to (010) usually show no twinning but occasionally have pericline twinning. They usually have the shape of Fig. 199*b*, page 230, with a fairly straight edge parallel to the *c*-axis, which is due to imperfect

FIG. 215.—(×9) Combined Carlsbad-albite twinning in plagioclase. (× nicols.)

FIG. 216.—(×12) Albite, Carlsbad, and pericline twinning in plagioclase. (× nicols.)

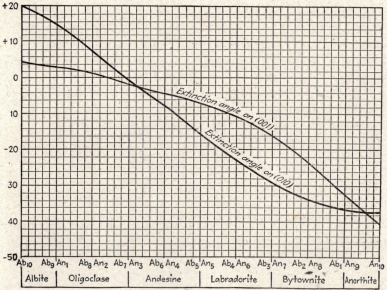

FIG. 217.—Curves showing the extinction angles of oriented sections of the plagioclase feldspars parallel to (001) and (010). Schuster's method.

cleavage parallel to {110} and {1$\bar{1}$0}. Extinction angles for sections parallel to (001) and (010) are given in the curves of Fig. 217, above. More accurate determinations for the sodic plagioclases can be made on (010) sections than on (001) sections, as may be seen on inspection of the curves.

MAXIMUM EXTINCTION ANGLES OF ALBITE TWINS IN SECTIONS NORMAL TO {010}
Figs. 218, 219

This, the statistical method of Michel-Lévy (1877), is perhaps the most useful single method of determining the character of the plagioclase. Sections normal to {010} may be recognized (1) by the sharpness of the lines separating the twin-lamellae on changing the focus slightly, (2) by the equal illumination when the lamellae are parallel to the vibration planes of the nicols, and

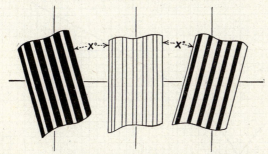

FIG. 218.—Diagram showing the method of determining the extinction angles in albite twins cut normal to (010) for the plagioclase feldspars.

(3) by the equality of the extinction angles on both sides of the twin lines.

Figure 218 shows the procedure. Many different sections cut normal to {010} can usually be found. The angles given on the curves (Fig. 219) are the maximum angles for the various kinds of plagioclase. It is necessary to determine the extinction angle in eight or ten different (or more if greater accuracy is desired) sections in the slide and use the maximum one. It is not necessary to have the extinction angles on each side exactly equal. The two angles may differ by as much as 5 or 6°, and in this case the average of the two readings is used. In recording the extinction angle the direction of the faster ray is used; otherwise no angle greater than 45° would ever be obtained. It will be noted that angles of $19\frac{1}{2}°$ or less appear twice on the curve. From Ab_{100} to $Ab_{79}An_{21}$ the angle is negative, and for those above

$Ab_{79}An_{21}$ it is positive. In the absence of the (001:100) edge—
and this is rarely present—positive and negative angles cannot be
distinguished. In order to distinguish plagioclase of the Ab_{100}
–$Ab_{79}An_{21}$ range from plagioclase of the $Ab_{79}An_{21}$–$Ab_{62}An_{38}$
range, indices of refraction or optical sign must be used. Most
of the first group have indices of refraction less than balsam and

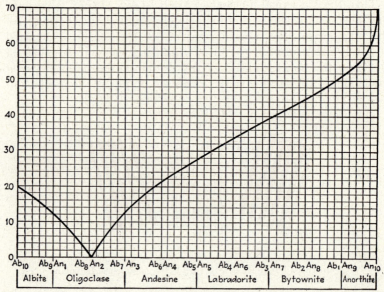

Fig. 219.—Curve showing the maximum extinction angle of albite twins cut
normal to (010) for the plagioclase feldspars. Michel-Lévy's method.

are optically positive. The others have indices of refraction
greater than balsam and are optically negative.

Extinction Angles of Combined Carlsbad and Albite Twins in Sections Normal to {010}
Fig. 221, Page 243

If both Carlsbad and albite twinning are present, a single section normal to (010) will suffice for the determination. Sections
twinned according to both the Carlsbad and albite laws will in
general appear as in Fig. 215, page 240, where four different
extinction positions for the crystal may be found. Now sections
normal to (010), the composition face for both kinds of twins,

may be recognized by the fact that in the 45° position the albite twinning disappears, and the crystal appears to be a simple Carlsbad twin. In the 0° position both the albite and Carlsbad twinning practically disappear. The extinction angles of the

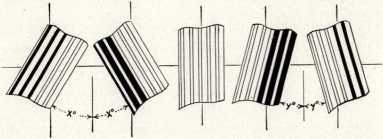

Fig. 220.—Diagram showing the method of determining the two sets of extinction angles ($x°$ and $y°$) in sections of combined Carlsbad-albite twins cut normal to (010).

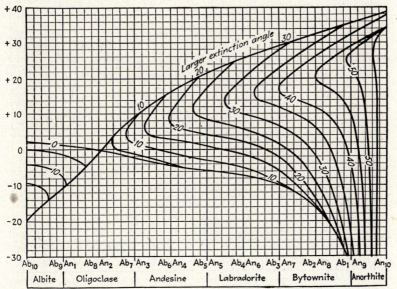

Fig. 221.—Curves showing extinction angles of combined Carlsbad-albite twins normal to (010) for the plagioclase feldspars. (*After F. E. Wright.*)

albite twins in each half of the section are measured. The procedure is shown in Fig. 220, above. The average of the two smaller is given on the horizontal lines of Fig. 221, and the average of the two larger on the curves. The intersection of these two lines gives a point that indicates the relative amounts of the albite and anorthite molecules.

For a plagioclase with the composition of about $Ab_{80}An_{20}$, Carlsbad twinning cannot be detected in thin sections. In this case the maximum extinction angle for albite twins in sections normal to (010) is 0° (see Fig. 219, page 242).

In general, the two sets of extinction angles will indicate two kinds of plagioclase, and it will be necessary to use some other method to distinguish them.

INDICES OF REFRACTION OF CLEAVAGE FLAKES
Fig. 222

The indices of refraction for plagioclase vary from $n_\beta = 1.529$ for pure albite to $n_\beta = 1.583$ for pure anorthite. The relief in

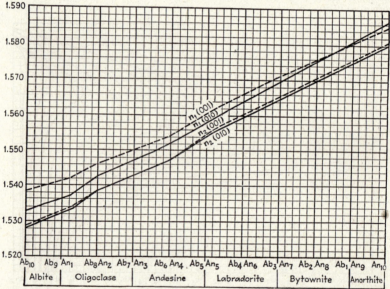

FIG. 222.—Curves showing the indices of refraction n_1 and n_2 of cleavage flakes of the plagioclase feldspars. (*After Tsuboi.*)

balsam then is low to rather low. The principal indices for the various plagioclases are listed under the individual descriptions.

It is rather difficult to determine the maximum and minimum indices unless special precautions are taken. Since these minerals are triclinic the ellipsoid axes are not in general parallel to the cleavages. In ordinary cleavage flakes special values of the indices of refraction n_1 and n_2 are obtained for the two orientations {001} and {010}. Curves for the indices are given in Fig. 222, page 244; the dotted lines are values for {001} and the solid lines for {010}. For these curves we are indebted to Tsuboi.

ANGLE OF THE RHOMBIC SECTION
Fig. 223

With pericline twinning the lamellae appear on the (010) cleavage face and make an angle with the (001:010) edge that

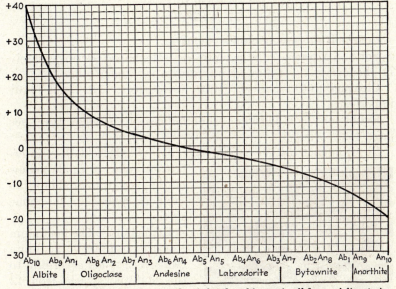

FIG. 223.—Curve showing the "angle of the rhombic section" for pericline twins of the plagioclase feldspars. (*After E. Schmidt.*)

varies with the character of the plagioclase. This is illustrated in Fig. 212, page 239, where σ is the angle in question. This angle is known as the *angle of the rhombic section*. The composition plane in a pericline twin is an irrational plane. Two individuals in a pericline twin are united in a plane, the intersection of which with faces (110), (1$\bar{1}$0), ($\bar{1}$10), and ($\bar{1}$10) is a rhombus; hence the term *rhombic section*. The curve of Fig. 223 shows the variation of this angle with the relative proportions of the albite and anorthite molecules. This method is about the only method of determining the kind of plagioclase by simple non-optical tests, but unfortunately pericline twinning is not always present. Here, as in all curves for plagioclase, positive angles are clockwise, and negative angles are counterclockwise. For plagioclase with the composition $Ab_{58}An_{42}$ the lamellae are exactly parallel to the edge ($\angle\alpha = 90°$, diclinic syngony).

ALBITE

Ab_{10} to Ab_9An_1 Triclinic

$$n_\alpha = 1.525 \text{ to } 1.532$$
$$n_\beta = 1.529 \text{ to } 1.536$$
$$n_\gamma = 1.536 \text{ to } 1.541$$
$$2V = 77 \text{ to } 82°; \text{ Opt. } (+)$$

Color.—Colorless in thin sections.

Form.—Albite occurs in plates or lath-shaped sections, rarely in phenocrysts. It may be intergrown with microcline.

Cleavage {001} perfect, {010} less perfect, {110} and {1$\bar{1}$0} imperfect.

Relief low, $n <$ balsam. (For indices of cleavage flakes see page 244.)

Birefringence rather weak, $n_\gamma - n_\alpha = 0.009$ to 0.011; interference colors are pale yellow of the first order, about the same as quartz in the same section.

Fig. 224.—($\times$25) Albite showing albite twinning.

Extinction.—The maximum extinction angle in albite twins (*i.e.*, twins according to the albite law) varies from 12 to 19°. In cleavage flakes parallel to (001) the extinction angle is 3° ±; on those parallel to (010), from 15 to 20°.

Twinning.—Polysynthetic twinning according to the albite law ({010} = twin-plane) is rarely absent. There may also be twinning according to the Carlsbad law (*c*-axis or [001] = twin-axis) either alone or combined with albite twinning. Pericline twinning (*b*-axis or [010] = twin-axis) is sometimes present.

Interference Figure.—The interference figure is biaxial positive with a large axial angle. Dispersion, $r < v$ weak.

Distinguishing Features.—Albite is distinguished from the other plagioclases by lower indices of refraction and by the various extinction angles.

Occurrence.—Albite occurs in some granites, in granite pegmatites, in veins, and in some metamorphic rocks. It is the only plagioclase that is at all common as a vein mineral. In some altered subsilicic igneous rocks (spilites) it is formed at the expense of calcic plagioclase as a deuteric mineral.

OLIGOCLASE

Ab_9An_1 to Ab_7An_3 Triclinic

$$n_\alpha = 1.532 \text{ to } 1.545$$
$$n_\beta = 1.536 \text{ to } 1.548$$
$$n_\gamma = 1.541 \text{ to } 1.552$$
$$2V = 82 \text{ to } 90°; \text{ Opt. } (+) \text{ or } (-)$$

Color.—Colorless in thin sections.

Form.—Oligoclase occurs in euhedral, subhedral, and anhedral crystals. The appearance is the same as for the other feldspars.

Cleavage {001} perfect, {010} less perfect, {110} and {1$\bar{1}$0} imperfect.

Relief low, n either less than, greater than, or about equal to that of balsam. (For indices of cleavage flakes see page 244.)

Birefringence weak or rather weak, $n_\gamma - n_\alpha = 0.007$ to 0.009; interference colors are gray or white of the first order.

Extinction.—The maximum extinction angle in albite twins (twinning according to the albite law) varies from 0 to 12°. The extinction angle on a (001) cleavage flake varies from 0 to 3°; on {010} flakes, from 0 to +15°.

Twinning.—As in albite.

Interference Figure.—The figure is biaxial, either positive or negative with a very large axial angle, or neutral (2V = 90°) for $Ab_{83}An_{17}$. Dispersion, $r > v$ weak.

Distinguishing Features.—Oligoclase is distinguished from the other plagioclases by extinction angles in twinned crystals and by the indices of refraction.

Occurrence.—Oligoclase is very common in persilicic igneous rocks such as granites and rhyolites; also in syenites, trachytes, and other igneous rocks. It is occasionally found in granite pegmatites and also in some metamorphic rocks.

ANDESINE

Ab_7An_3 to Ab_5An_5 Triclinic

$$n_\alpha = 1.545 \text{ to } 1.555$$
$$n_\beta = 1.548 \text{ to } 1.558$$
$$n_\gamma = 1.552 \text{ to } 1.562$$
$$2V = 76 \text{ to } 90°; \text{ Opt. } (+) \text{ or } (-)$$

Color.—Colorless in thin sections.

Form.—Andesine is found in euhedral to anhedral crystals.

Cleavage $\{001\}$ perfect, $\{010\}$ less perfect, $\{110\}$ and $\{1\bar{1}0\}$ imperfect.

Relief low, n always greater than balsam. (For indices of cleavage flakes see page 244.)

Birefringence weak, $n_\gamma - n_\alpha = 0.007$, so that the interference colors are gray or white of the first order.

Extinction.—The maximum extinction angle in albite twins (twins according to the albite law) varies from 13 to $27\frac{1}{2}°$. On (001) cleavage flakes the extinction angle varies from 0 to $-7°$; on (010) flakes, from 0 to $-16°$.

Twinning.—As in albite.

Interference Figure.—The figure is biaxial, either positive or negative with a large axial angle, or neutral ($2V = 90°$) for $Ab_{62}An_{38}$. Dispersion, $r < v$.

Distinguishing Features.—Andesine is distinguished from other plagioclases by maximum extinction angles of twinned crystals and by the indices of refraction.

Occurrence.—Andesine is a common and widely distributed mineral in igneous rocks of various types. It is especially common in diorites and andesites. Andesine also occurs in metamorphic rocks.

LABRADORITE

Ab$_5$An$_5$ to Ab$_3$An$_7$ Triclinic

$$n_\alpha = 1.555 \text{ to } 1.563$$
$$n_\beta = 1.558 \text{ to } 1.567$$
$$n_\gamma = 1.562 \text{ to } 1.571$$
$$2V = 76 \text{ to } 90°; \text{ Opt. } (+)$$

Color.—Colorless in thin sections, often with regularly arranged inclusions.

Form.—Labradorite occurs in euhedral to anhedral crystals. The anhedral crystals are often large as compared with those of other plagioclases.

Cleavage {001} perfect, {010} less perfect, {110} and {1$\bar{1}$0} imperfect.

Relief fairly low, $n >$ balsam. (For indices of cleavage flakes see page 244.)

Birefringence weak, $n_\gamma - n_\alpha = 0.007$ to 0.008; interference colors are gray or white of the first order.

Extinction.—The maximum extinction angle in albite twins (twinning according to the albite law) varies from

Fig. 225.—(×43) Plagioclase (labradorite) crystals in basalt.

27½ to 39°. The extinction angle on (001) cleavage flakes varies from -7 to $-16°$; on (010) flakes, from -16 to $-29°$.

Twinning.—As in albite.

Interference Figure.—The figure is usually biaxial positive with a large axial angle but is biaxial negative at times and neutral for Ab$_{32}$An$_{68}$. Dispersion, $r < v$.

Distinguishing Features.—Labradorite is distinguished from the other plagioclases by the maximum extinction angles of albite twins and by the indices of refraction.

Occurrence.—Labradorite is a very common mineral in subsilicic igneous rocks such as auganites, basalts, gabbros, and olivine gabbros. It is the principal constituent of most anorthosites. Labradorite also occurs in metamorphic rocks.

Bytownite

Ab$_3$An$_7$ to Ab$_1$An$_9$ Triclinic

$$n_\alpha = 1.563 \text{ to } 1.571$$
$$n_\beta = 1.567 \text{ to } 1.577$$
$$n_\gamma = 1.571 \text{ to } 1.582$$
$$2V = 79 \text{ to } 88°; \text{ Opt. } (-)$$

Color.—Colorless in thin sections.

Form.—Bytownite occurs in subhedral to anhedral crystals.

Cleavage {001} perfect, {010} less perfect, {110} and {1$\bar{1}$0} imperfect.

Relief moderate, $n >$ balsam.   (For indices of cleavage flakes see page 244.)

Birefringence rather weak, $n_\gamma - n_\alpha = 0.008$ to 0.011.   Interference colors are gray, white, or pale yellow of the first order.

Extinction.—The maximum extinction angle in albite twins (twinning according to the albite law) varies from 39 to 51°. The extinction angle on (001) cleavage flakes varies from -16 to $-32°$; on {010} flakes, from -29 to $-36°$.

Twinning.—As in albite.

Interference Figure.—The figure is biaxial negative with a very large axial angle.   Dispersion, $r > v$.

Distinguishing Features.—Bytownite is distinguished from other plagioclases by extinction angles and refractive indices.

Occurrence.—Bytownite usually occurs in gabbros, anorthosites, or basalts, but it is a comparatively rare mineral.

Anorthite

Ab$_1$An$_9$ to An$_{10}$ Triclinic

$$n_\alpha = 1.571 \text{ to } 1.575$$
$$n_\beta = 1.577 \text{ to } 1.583$$
$$n_\gamma = 1.582 \text{ to } 1.588$$
$$2V = 77 \text{ to } 79°; \text{ Opt. } (-)$$

Color.—Colorless.

Form.—Anorthite occurs in anhedral to subhedral plates or laths.

Cleavage {001} perfect, {010} less perfect, {110} and {1$\bar{1}$0} imperfect.

Relief fair, $n >$ balsam.   (For indices of cleavage flakes see page 244.)

Birefringence rather weak, $n_\gamma - n_\alpha = 0.011$ to 0.013; interference colors are gray, white, or yellow of the first order.

Extinction.—The maximum extinction angle in albite twins (twinning according to the albite law) varies from 51 to 70°. The extinction angle on (001) cleavage flakes varies from -32 to $-40°$; on (010) it is about $-37°$.

Twinning.—As in albite.

Interference Figure.—The figure is biaxial negative with a large axial angle.   Dispersion, $r > v$.

Distinguishing Features.—Anorthite is distinguished from other plagioclases by the extinction angles and refractive indices.

Occurrence.—Anorthite is rare compared with the other plagioclases.   It is found in a few contact-metamorphic deposits and in a few lavas.

THE FELDSPATHOIDS

The feldspathoids play the same role in some igneous rocks that the feldspars do.   They may either take the place of the feldspars or occur with them.   They are comparatively rare.

Analcime, though a zeolite, sometimes plays the part of feldspar in some igneous rocks.

THE FELDSPATHOIDS

Mineral	Chemical composition	Crystal system	Indices of refraction
Leucite...	KAl	Pseudo-isometric	$n_\alpha = 1.508,\ n_\gamma = 1.509$
Nepheline.	NaAl	Hexagonal	$n_\epsilon = 1.527-1.543,\ n_\omega = 1.530-1.547$
Cancrinite	NaAl + CO$_3$	Hexagonal	$n_\epsilon = 1.496-1.500,\ n_\omega = 1.507-1.524$
Sodalite...	NaAl + Cl	Isometric	$n\ \ = 1.483-1.487$
Haüyne...	NaAl + S	Isometric	$n\ \ = 1.496-1.510$
Melilite...	Ca,Mg,Al	Tetragonal	$n_\epsilon = 1.626-1.629,\ n_\omega = 1.632-1.634.$

Leucite

KAl(SiO$_3$)$_2$　　　　　　　　　　　　　　　　　　Pseudo-isometric

(Isometric above 600° C.)

$$n_\alpha = 1.508$$
$$n_\gamma = 1.509$$

Color.—Colorless in thin sections.

Form.—Leucite practically always occurs in euhedral crystals. The crystal form is the trapezohedron {211}, which shows octagonal sections. It often contains inclusions, and these may be arranged in a regular manner, either radially or concentrically.

Relief fair, n < balsam.

Birefringence very weak, $n_\gamma - n_\alpha = 0.001$; it is best detected by using the sensitive-violet test plate. Minute crystals may not show any birefringence.

Extinction is often wavy.

Twinning.—A characteristic feature of leucite is the complicated polysynthetic twinning in several directions, which often resembles that of microcline. When heated to about 600° C. the twinning disappears, which proves that KAl(SiO$_3$)$_2$ is dimorphous.

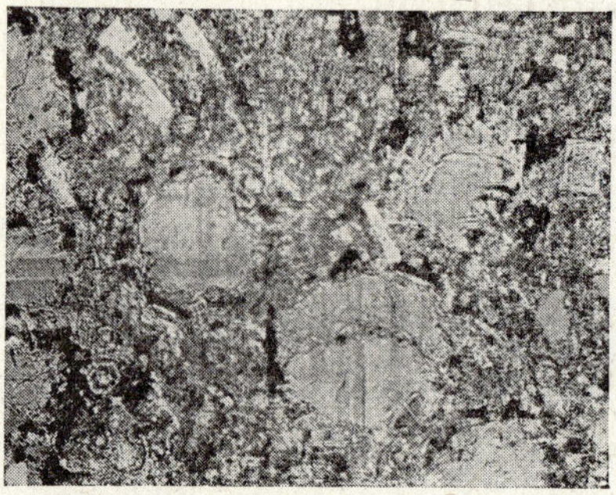

Fig. 226.—(×55) Leucite phenocrysts showing polysynthetic twinning with nepheline microphenocrysts in leucite phonolite.

Distinguishing Features.—Leucite resembles analcime. The latter shows weak birefringence but does not have definite polysynthetic twinning. Microcline has greater birefringence and lower relief.

Occurrence.—Leucite occurs almost exclusively as phenocrysts in lavas (leucite tephrite, leucitite, leucite basalt, leucite phonolite, etc.) and the corresponding tuffs. Leucite-bearing rocks are common in Italy but rare in most other parts of the world. A prominent American locality for leucite-bearing rocks is the Leucite Hills, Wyoming.

Leucite is exceedingly rare in grained igneous rocks (fergusite). In such rocks it usually has been altered to so-called *pseudoleucite* (orthoclase-nepheline mixture).

Nepheline

$(Na,K)(Al,Si)_2O_4$ (Eleolite in part)   Hexagonal
(Essentially $NaAlSiO_4$ (Hexagonal Subsystem)
with an excess of SiO_2)

$$n_\epsilon = 1.527 \text{ to } 1.543$$
$$n_\omega = 1.530 \text{ to } 1.547$$
$$\text{Opt. } (-)$$

Color.—Colorless to turbid in thin sections. It may show rows of inclusions.

Form.—Nepheline occurs in short prismatic hexagonal crystals (phenocrysts) in dense rocks and in anhedra in grained rocks. The crystals have rectangular and hexagonal sections and sometimes show zonal structure.

Cleavage imperfect parallel to $\{10\bar{1}0\}$, not always apparent.

Relief very low, n about the same as balsam but usually slightly higher.

Birefringence weak, $n_\omega - n_\epsilon = 0.003$ to 0.004;

Fig. 227.—($\times 15$) Nepheline with microcline in nepheline syenite. ($\times$ nicols.)

interference colors are gray of the first order.

Extinction parallel for rectangular sections. Basal sections are dark between crossed nicols.

Orientation.—The rectangular sections are length-fast.

Interference Figure.—Basal sections give a negative uniaxial figure without any rings.

Distinguishing Features.—The mineral most likely to be mistaken for nepheline is orthoclase, but the latter has better cleavage and is biaxial. It also resembles melilite and scapolite; the former has higher relief and the latter stronger birefringence.

Alteration.—Nepheline alters very readily to zeolites, sodalite, muscovite (gieseckite), cancrinite, or hydronephelite (a variety of natrolite).

Occurrence.—Nepheline is a rare mineral confined to soda-rich igneous rocks such as nepheline syenites, phonolites, and a few basaltic rocks. It never occurs with original quartz.

Cancrinite

$3NaAlSiO_4.CaCO_3.H_2O(?)$ Hexagonal

$$n_\epsilon = 1.496 \text{ to } 1.500$$
$$n_\omega = 1.507 \text{ to } 1.524$$
$$\text{Opt. } (-)$$

Color.—Colorless in thin sections to very pale yellow in sections a little thicker than normal.

Form.—Crystals are usually anhedral with a tendency toward elongation parallel to the *c*-axis. Euhedral crystals are rare.

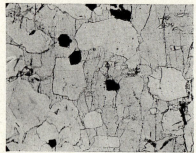

Fig. 228.—(×18) Cancrinite (gray) in nepheline syenite.

Cleavage.—Good cleavage parallel to $\{10\bar{1}0\}$.

Relief fair, $n <$ balsam.

Birefringence variable from rather weak (0.007) to moderate (0.028); the interference colors vary from first-order pale yellow up to middle second order.

Extinction parallel to outlines and to the cleavage traces. Some sections are isotropic.

Orientation.—Crystal outlines and cleavage traces are length-fast.

Interference Figure.—Basal sections give a negative uniaxial interference figure with not more than one or two rings.

Distinguishing Features.—Cancrinite is distinguished from similar minerals by its stronger birefringence.

Alteration.—There is sometimes alteration along cleavage cracks and fractures.

Related Minerals.—Hydronephelite, a zeolitic alteration of nepheline, is a closely related mineral.

Occurrence.—Cancrinite is a rare, but widely distributed, mineral characteristic of nepheline syenites. It is probably a deuteric mineral since it often surrounds and apparently replaces feldspars (see Fig. 228). Its associates are plagioclase (especially albite), microcline, nepheline, and sodalite.

Sodalite

3NaAlSiO$_4$.NaCl   Isometric

$$n = 1.483 \text{ to } 1.487$$

Color.—In thin sections colorless to gray, often with dark borders.

Form.—Sodalite occurs in six-sided euhedral crystals (cross sections of dodecahedra) and in anhedra.

Cleavage imperfect parallel to {110}, more likely to show on edges of the slide.

Relief fair, $n <$ balsam.

Birefringence nil.

Extinction.—Dark between crossed nicols.

Fig. 229.—($\times$17) Sodalite with feldspar in sodalite syenite.

Distinguishing Features.—Sodalite resembles analcime, but the latter is usually secondary. It may be necessary to make microchemical tests to confirm the determination of sodalite.

Alteration.—Sodalite is readily altered to zeolites.

Related Minerals.—Nosean (sometimes called *noselite*) is a related mineral containing the sulfate radical in the place of chlorin.

Occurrence.—Sodalite is practically confined to soda-rich igneous rocks such as

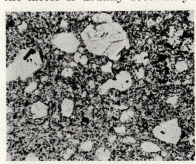

Fig. 230.—($\times$15) Sodalite phenocrysts in a porphyritic igneous rock.

syenites and trachytes. It is especially prominent in rocks called sodalite syenites. It is a common associate of nepheline.

Haüyne

$m3NaAlSiO_4.CaSO_4$ (Lazurite in part)   Isometric
$n3NaAlSiO_4.Na_2S$

$$n = 1.496 \text{ to } 1.510$$

With haüyne, often given in the less euphonious variant haüynite, is included lazurite, here considered to be simply a sulfid-bearing haüyne. (The name *lazurite* is discarded because of its similarity to lazulite.)

Fig. 231.—($\times$12) Haüyne with interstitial calcite.

Color.—Colorless, gray, pale blue, bluish green to deep blue in thin sections. The color may vary within a single crystal. Transparent to translucent.

Form.—Haüyne usually occurs in euhedral to anhedral crystals and in crystal aggregates. Both octahedrons and dodecahedrons are common crystal forms.

Cleavage.—It may show imperfect cleavage (dodecahedral, as in sodalite).

Relief.—Rather low, $n <$ balsam.

Birefringence.—Haüyne is usually isotropic, but occasionally it may show very weak birefringence up to about 0.004.

Distinguishing Features.—Haüyne resembles sodalite, but its refractive index is higher and its cleavage less prominent than that of soda-

Fig. 232.—($\times$23) Haüyne (dark gray) in lapis lazuli with pyrite, diopside, and muscovite.

lite. The presence of pyrite is characteristic of the sulfid-bearing haüyne of lapis lazuli.

Related Minerals.—Nosean (or noselite), a mineral of the sodalite group, is similar to haüyne, but it contains little or no calcium.

Occurrence.—Haüyne occurs in (1) soda-rich volcanic rocks such as phonolite and in (2) the contact-metamorphic limestones or muscovite-diopside gneisses known as lapis lazuli. Pyrite is an invariable constituent of lapis lazuli.

Melilite

$m(Ca_2Al_2SiO_7)$ Tetragonal
$n(Ca_2MgSi_2O_7)$

$$n_\epsilon = 1.626 \text{ to } 1.629$$
$$n_\omega = 1.632 \text{ to } 1.634$$
$$\text{Opt. } (-)$$

Melilite, although complex in composition, is essentially an isomorphous mixture of the two end members gehlenite (CaAl) and åkermanite (CaMg) according to Buddington.

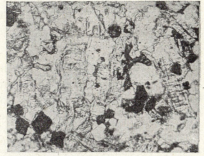

Color.—Colorless to pale yellow in thin sections.

Form.—The usual forms of melilite are euhedral crystals of tabular habit that show as rectangular sections. It often has "peg structure" due to lines normal to the length of the sections.

Fig. 233.—(×100) Melilite (fair relief) in nephelinite.

Cleavage indistinct parallel to {001}, which often appears as a single crack in the center of the section.

Relief fairly high, $n >$ balsam.

Birefringence weak, $n_\omega - n_\epsilon = 0.005$ to 0.006; interference colors are first-order gray and often anomalous Berlin blue.

Extinction parallel.

Orientation.—The rectangular sections are length-slow, since the mineral is tabular in habit and optically negative.

Interference Figure.—Basal sections give a uniaxial negative figure without any rings.

Distinguishing Features.—Elongated sections with weak birefringence and peg structure are characteristic. It somewhat resembles nepheline but the relief is fairly high instead of low.

Related Minerals.—Other minerals of the melilite group are gehlenite, $Ca_2Al_2SiO_7$, found in metamorphic limestones and åkermanite, $Ca_2MgSi_2O_7$, found in furnace slags and as a laboratory product.

Alteration.—Incipient alteration takes place along lines normal to the length of the crystal. This gives the so-called *peg structure*. It may also be altered to calcite and zeolites.

Occurrence.—Melilite occurs in subsilicic igneous rocks such as nepheline- and leucite-bearing lavas and in melilite basalts (alnöites). Usual associates are augite, olivine, nepheline, leucite, and perovskite. It is also a prominent constituent of a coarse-grained alkaline igneous rock called uncompahgrite in the San Juan region, Colorado.

Melilite is also found in furnace and in Portland-cement clinker slags.

THE PYROXENES

The pyroxenes include some of the most important rock-forming minerals. There are two subgroups, the orthorhombic orthopyroxenes and the monoclinic clinopyroxenes. Figure 234 shows typical cross sections with the angle (110:1$\bar{1}$0) about 93°. The cleavage is parallel to {110}, as shown in Fig. 234. There is sometimes parting parallel to {001} or {100}. Twinning

THE PYROXENES

	Mineral	Chemical composition	$n\alpha$	$n\beta$	$n\gamma$	2V	c: γ or Z
Orthorhombic	**Orthopyroxenes**						
	Enstatite...........	Mg	1.650 1.665	1.653 1.670	1.658 1.674	58–80°	0°
	Hypersthene.........	Mg,Fe	1.673 1.715	1.678 1.728	1.683 1.731	63–90°	0°
Monoclinic	**Clinopyroxenes**						
	Diopside............	Ca(Mg,Fe)	1.650 1.698	1.657 1.706	1.681 1.727	58–60°	37–44°
	Augite..............	Ca(Mg,Fe) + Al	1.688 1.712	1.701 1.717	1.713 1.737	58–62°	45–54°
	Pigeonite...........	CaMgFe	1.680 1.718	1.698 1.725	1.719 1.744	0–40°	22–45°
	Hedenbergite........	CaFe	1.732 1.739	1.737 1.745	1.751 1.757	60°	48°
	Aegirine-augite.......	NaCaMgFe	1.680 1.745	1.687 1.770	1.709 1.782	60°	52–75°
	Aegirine............	NaFe	1.745 1.777	1.770 1.823	1.782 1.836	60–66°	82–88°
	Jadeite.............	NaAl	1.655 1.666	1.659 1.674	1.667 1.688	70–75°	30–36°

with {100} as twin-plane is rather common. The minerals are all biaxial with rather large axial angles. The axial plane of the optic axes is (010).

One of the best means of identifying individual members of the group is by use of the maximum extinction angles in longitudinal sections.

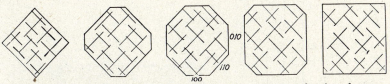

Fig. 234.—Cross sections of minerals of the pyroxene group showing cleavage.

Augite and pigeonite are by far the most common minerals of the group. The last three minerals of the tabulated list are known as *soda pyroxenes*.

THE ORTHOPYROXENES

Enstatite

MgSiO₃   (inc. Bronzite)   Orthorhombic

$$n_\alpha = 1.650 \text{ to } 1.665$$
$$n_\beta = 1.653 \text{ to } 1.670$$
$$n_\gamma = 1.658 \text{ to } 1.674$$
$$2V = 58 \text{ to } 80°; \text{ Opt. } (+)$$
$$a = \alpha \text{ or } X, b = \beta \text{ or } Y, c = \gamma \text{ or } Z$$

Color.—Colorless to neutral in thin sections. Bronzite has faint pleochroism.

Form.—Enstatite is found in prismatic crystals with the characteristic pyroxene cross section. Inclusions are common and produce what is known as *schiller* structure in the ferroan variety known as *bronzite*.

Cleavage {110} in two directions at nearly right angles (88 and 92°). Cleavage or parting parallel to {010} is also sometimes present. In longitudinal sections the cleavage traces are in one direction parallel to the outlines.

Relief high, $n >$ balsam.

Birefringence rather weak, $n_\gamma - n_\alpha = 0.008$ to 0.009; the maximum interference color is pale yellow of the first order.

Extinction parallel in most sections.

Twinning rarely present.

Orientation.—The crystals and cleavage traces are length-slow.

Interference Figure.—The figure is biaxial positive with a moderate to very large axial angle. The axial plane is {010}. Dispersion, $r < v$ weak.

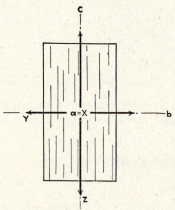

FIG. 235.—Orientation diagram of enstatite. Section parallel to (100).

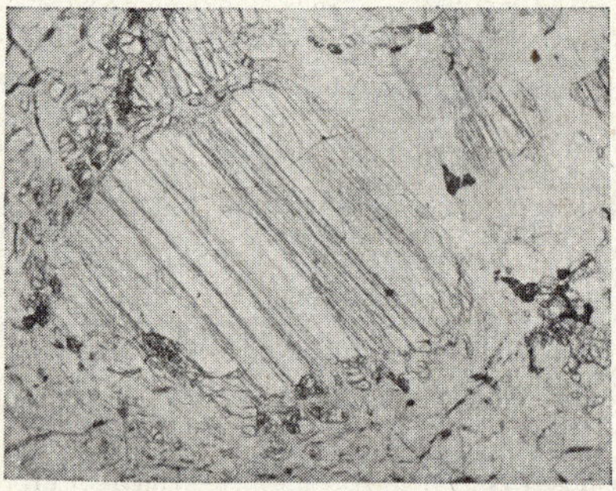

FIG. 236.—(×12) Enstatite in altered peridotite.

Intergrowth.—The intergrowth of enstatite with a monoclinic pyroxene is rather common. They have their c-axes in common and at first glance resemble polysynthetic twins.

Distinguishing Features.—Enstatite is distinguished from hypersthene by lack of pleochroism and from the monoclinic pyroxenes by parallel extinction.

Alteration.—It is common to find enstatite more or less altered to antigorite. Pseudomorphs of antigorite after enstatite are known as *bastite*.

Occurrence.—Enstatite is a characteristic mineral of sub-silicic igneous rocks and derived serpentites. It is also found in meteorites.

Hypersthene

$(Mg,Fe)SiO_3$ Orthorhombic

$$n_\alpha = 1.673 \text{ to } 1.715$$
$$n_\beta = 1.678 \text{ to } 1.728$$
$$n_\gamma = 1.683 \text{ to } 1.731$$
$$2V = 63 \text{ to } 90°; \text{ Opt. } (-)$$
$$a = \alpha \text{ or } X, \; b = \beta \text{ or } Y, \; c = \gamma \text{ or } Z$$

Color.—Neutral to pale green or pale red in thin sections. Pleochroic from greenish to pale reddish. Inclusions are common and produce schiller structure.

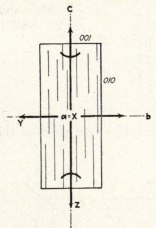

FIG. 237.—Orientation diagram of hypersthene. Section parallel to (100).

FIG. 238.—($\times 9$) Hypersthene (gray) in norite. The light mineral is plagioclase and the black one magnetite.

Form.—Hypersthene usually occurs in subhedral crystals of prismatic habit. The cross sections are nearly square.

Cleavage parallel to {110}; sometimes parallel to {010} and {100}.

Relief high, $n >$ balsam.

Birefringence rather weak, $n_\gamma - n_\alpha = 0.010$ to 0.016; the maximum interference color is yellow to red of the first order.

Extinction parallel in most sections.

Orientation.—The cleavage traces are length-slow.

Interference Figure.—The figure is biaxial negative with a large axial angle. The axial plane is {010}. Dispersion, $r > v$ weak.

Distinguishing Features.—The pleochroism is the most distinctive feature of hypersthene. It resembles some varieties of andalusite, but the latter mineral is length-fast, whereas hypersthene is length-slow.

Occurrence.—Hypersthene is found in a number of igneous rocks but is especially characteristic of norite, hypersthene gabbro, some andesite, and a peculiar hypersthene granite known as charnockite.

THE CLINOPYROXENES

DIOPSIDE

$Ca(Mg,Fe)(SiO_3)_2$

Monoclinic
$\angle\beta = 74°10'$

$$n_\alpha = 1.650 \text{ to } 1.698$$
$$n_\beta = 1.657 \text{ to } 1.706$$
$$n_\gamma = 1.681 \text{ to } 1.727$$
$$2V = 58 \text{ to } 60°; \text{ Opt. } (+)$$
$$b = \beta \text{ or } Y, c \wedge \gamma \text{ or } Z = -37 \text{ to } -44°.$$

Color.—Colorless, neutral, pale green to bright green in thin sections.

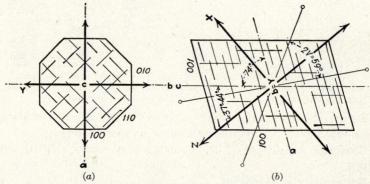

(a) (b)

FIGS. 239 a,b.—Orientation diagrams of diopside. Sections (a) normal to the c-axis and (b) parallel to (010).

Form.—Diopside usually occurs in subhedral crystals of short prismatic habit. Cross sections are four or eight sided.

Cleavage parallel to {110} and so in two directions at angles of 87 and 93°. Parting parallel to {001} is sometimes developed.

Relief fairly high, $n >$ balsam.

Birefringence rather strong, $n_\gamma - n_\alpha = 0.029$ to 0.031; the maximum interference color is about upper second order.

Extinction.—The maximum extinction angle in sections cut parallel to the *c*-axis varies from -37 to $-44°$. In cross sections the extinction is symmetrical to the cleavage traces.

Orientation.—The extinction direction that makes the smaller angle with the cleavage traces in longitudinal sections is the slower ray.

Twinning.—Twins with {100} as twin-plane are rather common. Polysynthetic twinning with {001} as twin-plane is common as secondary twinning.

Fig. 240.—(×36) Diopside from contact-metamorphic zone.

Interference Figure.—Diopside gives a biaxial positive figure with a rather large axial angle. The axial plane is {010}. Dispersion, $r < v$ weak. Flakes parallel to the {001} parting give a good optic-axis figure.

Distinguishing Features.—Diopside is distinguished from hedenbergite by lower refractive indices. From tremolite it is distinguished by larger extinction angle. Augite has a little higher extinction angle ($c \wedge \gamma$ or Z) and is usually a darker color. Pigeonite has a smaller axial angle.

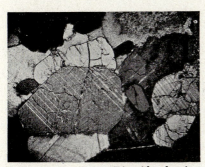

Fig. 241.—(×12) Diopside showing polysynthetic twinning. The section left of the center is cut approximately normal to [001]. (× nicols.)

Alteration.—Diopside is sometimes more or less altered to tremolite-actinolite.

Related Minerals.—Chrome diopside and omphacite are similar to diopside.

Occurrence.—Diopside is especially characteristic of contact-metamorphic zones. It occurs with garnet, wollastonite, idocrase, and other silicates. It is also found in some gneisses and schists and in some igneous rocks.

DIALLAGE

This name is used for a variety of ferroan diopside with prominent parting parallel to {100}. It is scratched by a knife blade and has a pearly, more or less metalloidal luster. Parting flakes have parallel extinction, but sections cut parallel to {010} give the large extinction angles characteristic of diopside. The parting flakes give an uncentered optic-axis interference figure that serves to distinguish it from the orthopyroxenes. Diallage is especially characteristic of coarse-grained gabbros.

AUGITE

$Ca(Mg,Fe)(SiO_3)_2[(Al,Fe)_2O_3]_x$ Monoclinic
$$\angle\beta = 74°10'$$

$$n_\alpha = 1.688 \text{ to } 1.712$$
$$n_\beta = 1.701 \text{ to } 1.717$$
$$n_\gamma = 1.713 \text{ to } 1.737$$
$$2V = 58 \text{ to } 62°; \text{ Opt. } (+)$$
$$b = \beta \text{ or } Y, c \wedge \gamma \text{ or } Z = -45 \text{ to } -54°$$

Color.—Almost colorless, neutral, pale greenish, or pale purplish brown in thin sections. Zonal structure is sometimes

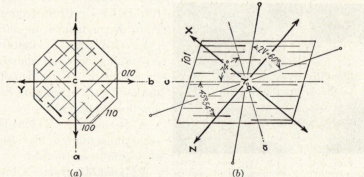

(a)	(b)

FIGS. 242 *a,b.*—Orientation diagrams of augite. Sections (*a*) normal to the c-axis and (*b*) parallel to (010).

present. Pleochroism absent to weak; it is best shown in (100) sections.

Form.—Augite usually occurs in short prismatic crystals with four- or eight-sided cross sections.

Cleavage {110} in two directions at angles of 87 and 93°. Cleavage traces are in one direction in longitudinal sections. Diallage has prominent parting parallel to {100}.

Relief high, $n >$ balsam.

Birefringence moderate, $n_\gamma - n_\alpha = 0.021$ to 0.025. The maximum interference color is about middle second order. Sections parallel to {100} have low first-order colors.

Fig. 243.—(×12) Augite (high relief) with plagioclase in basalt.

Extinction.—The maximum extinction angle of longitudinal sections varies from 36 to 45°. These sections have the maximum interference colors for the slide. Some varieties have a peculiar concentric wavy extinction known as the *hourglass* structure. Cross sections have parallel or symmetrical extinction depending upon whether {100} and {010} or {110} predominates.

Orientation.—The extinction direction that makes the smaller angle with the cleavage traces is the faster ray.

Twinning.—Twins with {100} as twin-plane are common; these often appear as twin seams. Polysynthetic twins with {001} as twin-plane are occasionally found. Combined {100} twins with {001} polysynthetic twins give what is known as *herringbone* structure.

Interference Figure.—The figure is biaxial positive with a rather large axial angle. The axial plane is {010}. Dispersion, $r > v$.

Distinguishing Features.—Augite is often difficult to distinguish from diopside. The extinction angle $c \wedge \gamma$ or Z is a little smaller and the color lighter in diopside.

Alteration.—There are two common alteration products of augite: (1) hornblende formed at a late magmatic stage and in parallel position on the augite (see Fig. 266, page 287); (2) uralite or secondary tremolite-actinolite formed by hydrothermal alteration.

Occurrence.—Augite is a very common mineral in subsilicic igneous rocks such as auganites, gabbros, basalts, olivine gabbros, limburgites, and peridotites.   Locally it is found in gneisses and granulites.

Augite is also a common detrital mineral.

PIGEONITE

$mCaMg(SiO_3)_2$ (Enstatite-augite)   Monoclinic
$n(Mg,Fe)SiO_3$ $\angle\beta = (?)$

$$n_\alpha = 1.680 \text{ to } 1.718$$
$$n_\beta = 1.698 \text{ to } 1.725$$
$$n_\gamma = 1.719 \text{ to } 1.744$$
$$2V = 0 \text{ to } 40°; \text{ Opt. } (+)$$
$$b = \beta, c \wedge \gamma \text{ or } Z = -22 \text{ to } -45°$$

Pigeonite (named by A. N. Winchell from Pigeon Point, Minn.) is an isomorphous mixture of diopside and clinoenstatite, a mono-

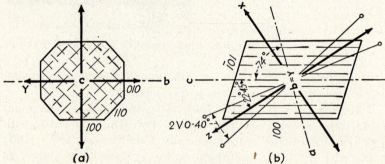

Figs. 244 *a,b.*—Orientation diagrams of pigeonite.   Sections (*a*) normal to the c-axis and (*b*) parallel to (010).

clinic pyroxene found in meteorites and also produced in the laboratory.

Color.—Colorless or neutral in thin sections.   It may show faint pleochroism.

Form.—Pigeonite usually occurs in anhedral crystals.

Cleavage in two directions {110} at angles of about 87 and 93°.

Relief fairly high, $n >$ balsam.

Birefringence moderate, $n_\gamma - n_\alpha = 0.021$ to 0.033; so the maximum interference color varies from lower to upper second order.

Extinction.—The maximum extinction angle varies from about 22 to 45°.   It increases with increase of the clinoenstatite content.

Orientation.—The extinction direction that makes the smaller angle with the cleavage traces in longitudinal sections is the slower ray.

Twinning.—Polysynthetic twinning with (100) as the twin-plane is characteristic of pigeonite.

Interference Figure.—The interference figure of pigeonite is biaxial positive with a rather small to very small axial angle. The axial plane is usually (010), but in varieties with very low calcium content the axial plane is normal to (010). For a certain composition pigeonite should have 2V = 0°; a uniaxial pigeonite from Mull has been described by Hallimond.

Distinguishing Features.—The only mineral that is likely to be mistaken for pigeonite is augite, from which it may be distinguished by its small axial angle.

Related Minerals.—Clinoenstatite, the calcium-free end member of the pigeonite series, is a well-known laboratory product (the Geophysical Laboratory), but as a mineral it is known only in meteorites.

Occurrence.—According to Barth, pigeonite is the most abundant member of the pyroxene group in volcanic rocks. It occurs in basalts, dolerites, and diabases. It is largely confined to the groundmass and is rarely found in phenocrysts.

Hedenbergite

$Ca(Fe,Mg)(SiO_3)_2$ Monoclinic
$\angle\beta = 74°30'$

$$n_\alpha = 1.732 \text{ to } 1.739$$
$$n_\beta = 1.737 \text{ to } 1.745$$
$$n_\gamma = 1.751 \text{ to } 1.757$$
$$2V = 60°; \text{ Opt. } (+)$$
$$b = \beta \text{ or } Y, c \wedge \gamma \text{ or } Z = -48°$$

Color.—Neutral to greenish in thin sections.

Form.—Hedenbergite usually occurs in columnar aggregates.

Cleavage {110} in two directions at angles of 87 and 93° (like the other pyroxenes).

Relief very high, $n >$ balsam.

Birefringence moderate, $n_\gamma - n_\alpha = 0.018$ to 0.019; the maximum interference color is about first-order violet.

Extinction.—The maximum extinction angle in longitudinal sections is about 42° ($c \wedge \alpha$ or X).

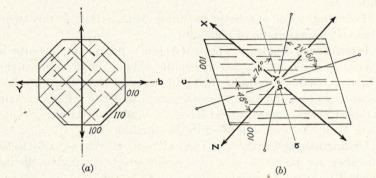

(a)

(b)

Figs. 245 *a,b*.—Orientation diagrams of hedenbergite. Sections (*a*) normal to the *c*-axis and (*b*) parallel to (010).

Orientation.—The extinction direction that makes the smaller angle with the cleavage traces is the faster ray.

Fig. 246.—(×13) Hedenbergite from contact-metamorphic zone.

Interference Figure.—The figure is biaxial positive with a rather large axial angle. The axial plane is {010}. Dispersion, $r > v$ weak.

Distinguishing Features.—Hedenbergite is distinguished from diopside and augite by higher indices of refraction.

Occurrence.—The characteristic occurrence of hedenbergite is in contact-metamorphic zones. It is often associated with iron ores as a skarn mineral.

Aegirine-Augite

Intermediate between
aegirine and augite
in chemical composition

Monoclinic

$$n_\alpha = 1.680 \text{ to } 1.745$$
$$n_\beta = 1.687 \text{ to } 1.770$$
$$n_\gamma = 1.709 \text{ to } 1.782$$
$$2V = ca.\ 60°;\ \text{Opt.}\ (+)\ \text{or}\ (-)$$
$$b = \beta \text{ or } Y,\ c \wedge \alpha \text{ or } X = -15 \text{ to } -38°$$

Color.—Green in thin sections.   Pleochroic from yellow-green (β or Y) to greenish (α or X, γ or Z).

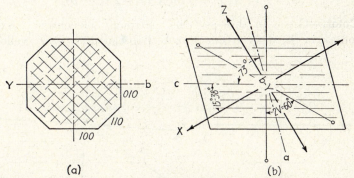

(a)   (b)

FIGS. 247 *a,b.*—Orientation diagrams of aegirine-augite.   Sections (*a*) normal to
the *c*-axis and (*b*) parallel to (010).

Form.—Aegirine-augite usually occurs in euhedral crystals of short prismatic habit with {100} as the dominant form.

Cleavage in two directions {110} at angles of 87 and 93°.

Relief high, $n >$ balsam.

Birefringence rather strong, $n_\gamma - n_\alpha = 0.029$ to 0.037; interference colors range up to the middle of the second order.

Extinction.—The maximum extinction in longitudinal sections varies from -15 to $-38°$.

Orientation.—In sections with the maximum extinction angle the extinction direction nearest the *c*-axis is the faster ray.

Twinning.—Twins with {100} as twin-plane are common.

Interference figure is biaxial positive with a rather large axial angle.   The axial plane is {010}.   Dispersion, $r > v$.

Distinguishing Features.—Aegirine-augite resembles aegirine but may be distinguished by the larger extinction angles.   It is most easily distinguished from the green varieties of hornblende by pyroxene cross sections and cleavage.

Occurrence.—Aegirine-augite occurs in soda-rich igneous rocks such as syenites, trachytes, nepheline syenites, phonolites, etc.

Aegirine

$NaFe(SiO_3)_2$

Monoclinic
$\angle \beta = 73°9'$

$$n_\alpha = 1.745 \text{ to } 1.777$$
$$n_\beta = 1.770 \text{ to } 1.823$$
$$n_\gamma = 1.782 \text{ to } 1.836$$
$$2V = 60 \text{ to } 66°; \text{ Opt. } (-)$$
$$b = \beta \text{ or } Y, c \wedge \alpha \text{ or } X = -2 \text{ to } -10°$$

Color.—Green in thin sections. Strongly pleochroic. Axial colors: α or X, dark green; β or Y, light green; γ or Z, yellow.

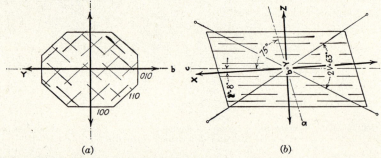

(a)   (b)

FIGS. 248 *a,b.*—Orientation diagrams of aegirine. Sections (*a*) normal to the *c*-axis and (*b*) parallel to (010).

Form.—Aegirine is usually found in long prismatic crystals, often bladed, with the typical four- to eight-sided cross section of the pyroxenes but with {100} frequently better developed than {010}.

Cleavage {110} in two directions at angles of 87 and 93°.

Relief high, $n >$ balsam.

Birefringence strong to very strong, $n_\gamma - n_\alpha = 0.037$ to 0.059; the interference colors should be third or fourth order but may be difficult to determine because the color of the mineral may mask the interference colors.

Extinction.—The maximum extinction angle in longitudinal sections is very small (from 2 to 10°).

Orientation.—The crystals are always length-fast.

Interference Figure.—The figure is biaxial negative with a rather large axial angle. The axial plane is {010}. Dispersion, $r > v$.

Distinguishing Features.—Aegirine resembles some of the amphiboles but is distinguished by the small maximum extinction angle and length-fast character. All the other monoclinic pyroxenes have larger extinction angles.

Related Minerals.—Acmite is a pyroxene closely related to aegirine. It differs from the latter in its brown color.

Occurrence.—Aegirine, although a rather rare mineral,

FIG. 249.—(×34) Aegirine rims around aegirine-augite in nepheline syenite.

is characteristic of soda-rich igneous rocks such as nepheline syenite, phonolite, syenite, trachyte, soda granite, soda aplite, etc. In these rocks it often occurs as an overgrowth on aegirine-augite crystals.

Jadeite

$NaAl(SiO_3)_2$ (Jade in part) Monoclinic
$\angle \beta = 72°44\frac{1}{2}'$

$$n_\alpha = 1.655 \text{ to } 1.666$$
$$n_\beta = 1.659 \text{ to } 1.674$$
$$n_\gamma = 1.667 \text{ to } 1.688$$
$$2V = 70 \text{ to } 75°; \text{ Opt. } (+)$$
$$b = \beta \text{ or } Y, c \wedge \gamma \text{ or } Z = -30 \text{ to } -36°$$

Color.—Colorless to green in thin sections. Some of the deeply colored varieties are pleochroic.

Form.—Jadeite usually appears in granular to columnar or somewhat fibrous aggregates. The texture varies from fine-grained to coarse-grained. Euhedral crystals are exceedingly rare.

Cleavage {110} in two directions at angles of about 87 and 93°.

Relief rather high, $n >$ balsam.

Birefringence moderate, $n_\gamma - n_\alpha = 0.012$ to 0.023; the maximum interference colors are second order.

Extinction.—The maximum extinction angle in longitudinal sections varies from 30 to 44°.

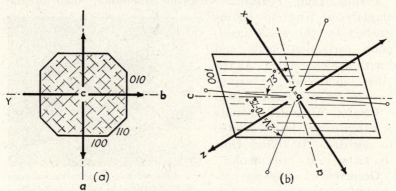

FIGS. 250 *a,b.*—Orientation diagrams of jadeite. Sections (*a*) normal to the *c*-axis and (*b*) parallel to (010).

Orientation.—The extinction direction nearest the *c*-axis is the slower ray.

Twinning.—Twins with {100} as twin-plane are occasionally found.

Interference Figure.—The figure is biaxial positive with a large axial angle. The axial plane is {010}. Dispersion, $r < v$.

FIG. 251.—(×25) Jadeite in jadeite rock.

Distinguishing Features.—Jadeite is distinguished from nephrite (variety of tremolite-actinolite) by larger extinction angle and higher refractive indices. From diopside it is distinguished by smaller maximum extinction angles and columnar habit.

Related Minerals.—Chloromelanite is an iron-bearing greenish black jadeite that is strongly pleochroic in thin sections.

Alteration.—Jadeite is sometimes found more or less altered to tremolite-actinolite.

Occurrence.—Jadeite occurs exclusively in jadeite rock (jadeitite) a monomineralic metamorphic rock formed according to Grubenmann in the deep zone of metamorphism. Albite is mentioned as one of the characteristic associates. The origin of jadeitite is obscure; it is found in only a few localities in Upper Burma, eastern Turkestan, northern Italy, and Guatemala.

JADE

The name *jade* (Chinese, *Yü*) is a general term for two distinct minerals: (1) nephrite, a tough, compact variety of tremolite-actinolite and (2) jadeite, an independent member of the pyroxene group, including its iron-bearing variety, chloromelanite.

The jadelike minerals or *pseudojades* include bowenite, a hard tough serpentine; californite, a compact variety of idocrase; "South African jade," a massive green grossularite; "Oregon jade," also a variety of grossularite; "Styrian jade," a kind of pseudophite, a compact chlorite; as well as sillimanite, pectolite, and wollastonite.

Spodumene

$LiAl(SiO_3)_2$ Monoclinic
$$\angle\beta = 69°40'$$

$$n_\alpha = 1.651 \text{ to } 1.668$$
$$n_\beta = 1.665 \text{ to } 1.675$$
$$n_\gamma = 1.677 \text{ to } 1.681$$
$$2V = 54 \text{ to } 69°; \text{ Opt. } (+)$$
$$b = \beta \text{ or } Y, c \wedge \gamma \text{ or } Z = -23 \text{ to } -27°$$

Color.—Colorless in thin sections. Some varieties show color (amethystine for kunzite, greenish for hiddenite) in thick sections and are pleochroic.

Form.—Spodumene usually occurs in euhedral crystals tabular parallel to {100} and elongated in the direction of [001]. Crystals as a rule are inclined to be large but are sometimes of the order of several millimeters.

Cleavage perfect parallel to {110} $(110 \wedge 1\overline{1}0) = 93°$. Parting parallel to {100}, which is often more prominent than the cleavage.

Relief fairly high, $n >$ balsam.

Birefringence moderate, $n_\gamma - n_\alpha = 0.013$ to 0.027; the maximum interference color varies from upper first order to middle second order.

Extinction.—The maximum extinction angle in longitudinal sections varies from 23 to 27°. In cross sections the extinction is

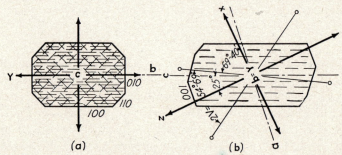

FIGS. 252 *a,b.*—Orientation diagram of spodumene. Sections (*a*) normal to the *c*-axis and (*b*) parallel to (010).

parallel or symmetrical. Oriented sections cut parallel to the (100) parting have parallel extinction.

Orientation.—The extinction direction that makes the smaller angle with the cleavage traces is the slower ray.

FIG. 253.—(×28) Spodumene (high relief) with quartz.

Twinning.—Twins with {100} as twin-plane are known.

Interference Figure.—Spodumene gives a positive biaxial interference figure with a rather large axial angle. The axial plane is (010). Dispersion, $r < v$.

Distinguishing Features.—Spodumene resembles diopside in general appearance, from which it may be distinguished by smaller extinction angle ($c\gamma = ca.$ 25°) and frequently by very conspicuous (100) parting.

Alteration.—Spodumene is sometimes altered to a mixture of albite and muscovite known as *cymatolite*. The muscovite here is an alteration of eucryptite (hexagonal $LiAlSiO_4$).

Kunzite is known to be altered to cookeite, a lithium aluminum silicate related to lepidolite.

Occurrence.—The typical occurrence of spodumene is the lithium granite pegmatites where it is associated with albite, lepidolite, elbaite, and rare lithium minerals.

THE PYROXENOIDS

Berman has suggested the term *pyroxenoid* for a number of pyroxene-like metasilicates that are not isomorphous with any of the pyroxenes. Pyroxenoids include rhodonite, bustamite, pectolite, and wollastonite. Of these, only wollastonite is important as a rock-forming mineral.

Wollastonite

$CaSiO_3$

Triclinic
$\angle \alpha = 90°0'$
$\angle \beta = 95°16'$
$\angle \gamma = 103°22'$

$$n_\alpha = 1.620$$
$$n_\beta = 1.632$$
$$n_\gamma = 1.634$$
$$2V = ca.\ 39°;\ Opt.\ (-)$$
b almost $\parallel$ to β or Y, $c \wedge \alpha$ or X $= +32°$

Color.—Colorless in thin sections.

Form.—Wollastonite usually appears in columnar or fibrous aggregates. The cross sections are nearly rectangular.

Cleavage in several directions in the zone [010]; perfect parallel to (100), less perfect (001) and ($\overline{1}$02) and imperfect (101) and ($\overline{1}$01).

Relief fairly high, $n >$ balsam.

Birefringence rather weak, $n_\gamma - n_\alpha = 0.014$; the maximum interference color is about orange of the first order. Longitudinal sections show gray or white interference colors.

Extinction parallel or almost parallel in longitudinal sections, oblique in cross sections.

Orientation.—Longitudinal sections are either length-slow or length-fast since the elongation is in the direction of the b-axis and $b = \beta$ or Y.

Twinning.—Twins with {100} as twin-plane are known.

Interference Figure.—The figure is biaxial negative with a moderate axial angle. Since the axial plane is almost parallel to

{010}, the figure lies normal to the length of the crystals. Dispersion, $r > v$ weak.

Distinguishing Features.—Tremolite greatly resembles wollastonite not only in hand specimens but in thin sections as well.

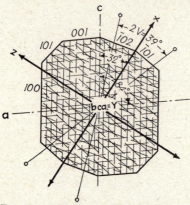

Tremolite, however, has oblique extinction and the typical amphibole cross section and cleavage. In tremolite the interference figure lies along the length of the crystal; in wollastonite it lies almost normal to the length.

Related Minerals.—CaSiO₃ is trimorphous. In addition to triclinic wollastonite, there are also *parawollastonite* (Peacock), which is monoclinic, and the high-temperature modification called pseudo-wollastonite (Rankin and Wright). The uniaxial pseudo-wollastonite has recently been described as a natural mineral in Persia (McLintock).

Fig. 254.—Orientation diagram of wollastonite. Section parallel to (010).

Pectolite [HNaCa₂(SiO₃)₃] is much like wollastonite in some of its properties.

Occurrence.—Wollastonite occurs in contact-metamorphic zones, in some schists and gneisses, and in limestone inclusions in volcanic rocks (parawollastonite).

THE AMPHIBOLES

The amphiboles have rhombic to pseudo-hexagonal cross sections and perfect cleavage parallel to {110} at angles of about

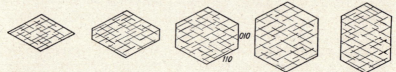

Fig. 255.—Cross sections of minerals of the amphibole group showing cleavage.

56 and 124°, as shown in Fig. 255. Twinning parallel to {100} is fairly common. With the exception of crossite, in all the amphiboles the plane of the optic axes is {010}.

AMPHIBOLE GROUP

	Mineral	Chemical composition	n_α	n_β	n_γ	2V	$c:\gamma$ or Z
Ortho-rhombic	Anthophyllite...........	MgFe	{ 1.598 { 1.652	1.615 1.662	1.623 } 1.676 }	70–90°	0°
Monoclinic	Cummingtonite.........	FeMg	{ 1.639 { 1.657	1.645 1.669	1.664 } 1.686 }	68–87°	15–20°
	Grunerite..............	Fe	{ 1.657 { 1.663	1.684 1.697	1.699 } 1.717 }	79–86°	10–14°
	Tremolite-actinolite (inc. nephrite).............	CaMgFe	{ 1.600 { 1.628	1.613 1.644	1.625 } 1.655 }	79–85°	10–20°
	Hornblende.............	CaMgFeAl	{ 1.614 { 1.675	1.618 1.691	1.633 } 1.701 }	52–85°	12–30°
	Lamprobolite............	CaMgFeAl	{ 1.670 { 1.692	1.683 1.730	1.693 } 1.760 }	64–80°	0–10°
	Riebeckite..............	NaFe	1.693	1.695	1.697	Large	85°
	Glaucophane............	NaAlFe	{ 1.621 { 1.655	1.638 1.664	1.639 } 1.668 }	45°	4– 6°

The amphibole group is more or less parallel to the pyroxene group. Corresponding members of the two groups, however, are not dimorphous. Hornblende is by far the most common mineral of the group. Cummingtonite is a pale brown monoclinic amphibole with the composition of anthophyllite. It is rare. Basaltic hornblende is considered a distinctive mineral under the name lamprobolite. Riebeckite, glaucophane, and a few rarer minerals are known as *soda amphiboles*.

The amphibole group is one of the most complex of all mineral groups. There are many amphiboles that cannot be placed under any of the minerals listed here.

Anthophyllite

$(Mg,Fe)_7(OH)_2(Si_4O_{11})_2$ Orthorhombic

$$n_\alpha = 1.598 \text{ to } 1.652$$
$$n_\beta = 1.615 \text{ to } 1.662$$
$$n_\gamma = 1.623 \text{ to } 1.676$$
$$2V = 70 \text{ to } 90°; \text{ Opt. } (+)$$
$$a = \alpha \text{ or } X, \ b = \beta \text{ or } Y, \ c = \gamma \text{ or } Z$$

Color.—Colorless or pale colored in thin sections. Some of the colored varieties show pleochroism.

Form.—Long prismatic crystals and columnar to fibrous aggregates are characteristic of anthophyllite. It is sometimes asbestiform.

Cleavage in two directions {110} at angles of 54 and 126°. Cross fractures are common.

Relief high, $n >$ balsam.

Birefringence moderate, $n_\gamma - n_\alpha = 0.016$ to 0.025; interference colors range up to low second order.

Extinction parallel in longitudinal sections, in cross sections symmetrical to outline or cleavage.

Orientation length-slow.

Twinning absent.

Interference Figure.—The figure is biaxial positive with a large axial angle, or neutral (2V = 90°). The axial plane is {010}. Dispersion, $r > v$ or $r < v$.

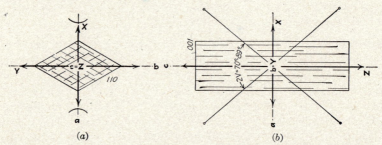

(a)   (b)

FIGS. 256 *a,b.*—Orientation diagrams of anthophyllite. Sections (*a*) normal to the *c*-axis and (*b*) parallel to (010).

Distinguishing Features.—Anthophyllite resembles tremolite-actinolite and also cummingtonite but may be distinguished from these minerals by its parallel extinction.

Related Minerals.—Gedrite is an aluminous variety of anthophyllite. It is optically negative instead of positive.

Alteration.—Anthophyllite is often altered to talc. The partially altered mineral was formerly called *hydrous anthophyllite.*

Occurrence.—Anthophyllite is characteristic of metamorphic rocks. It is the main constituent of anthophyllite schist and is also a secondary mineral in peridotites and dunites.

Anthophyllite is the principal constituent of mass-fiber asbestos.

Cummingtonite

$(Fe,Mg)_7(OH)_2(Si_4O_{11})_2$ Monoclinic

$$n_\alpha = 1.639 \text{ to } 1.657$$
$$n_\beta = 1.645 \text{ to } 1.669$$
$$n_\gamma = 1.664 \text{ to } 1.686$$
$$2V = 68 \text{ to } 87°; \text{ Opt. } (+)$$
$$b = \beta \text{ or } Y, c \wedge \gamma \text{ or } Z = -15 \text{ to } -20°$$

Color.—Colorless to neutral in thin sections. It may show slight pleochroism.

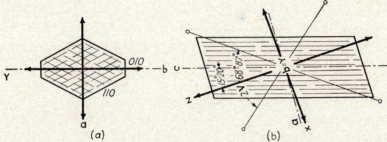

FIGS. 257 *a,b.*—Orientation diagrams of cummingtonite. Sections (*a*) normal to the *c*-axis and (*b*) parallel to (010).

Form.—Cummingtonite usually occurs in parallel to subradiating aggregates of prismatic crystals. The crystals are sometimes curved.

Cleavage.—Cleavage in two directions at angles of about 56 and 124° as in the other amphiboles.

Relief moderately high, $n >$ balsam.

Birefringence rather strong, $n_\gamma - n_\alpha = 0.025$ to 0.029.

Extinction.—The maximum extinction angle in longitudinal sections varies from about 15 to 20°; the angle increases with increase of magnesium content.

FIG. 258.—(×25) Cummingtonite in hornfels.

Orientation.—Elongate sections are length-slow.

Twinning.—Twins with {100} as twin-plane are highly characteristic of cummingtonite. The twinning is polysynthetic, and the twin-lamellae are usually very narrow.

Interference Figures.—The figure is biaxial positive with a large axial angle. The axial plane is (010). Dispersion, $r < v$.

Distinguishing Features.—Cummingtonite has a larger extinction angle and lower indices of refraction than grunerite. Cummingtonite is optically positive, and grunerite is negative. From tremolite, cummingtonite is distinguished by higher indices of refraction. Anthophyllite is very similar to cummingtonite but may be distinguished by its parallel extinction and absence of twinning.

Occurrence.—Cummingtonite, as far as known, is confined to metamorphic rocks. It is a characteristic mineral in schists at the Homestake Mine in South Dakota. The senior author has recently found it to be a characteristic mineral of hornfels at several localities in California.

Grunerite

$Fe_7(OH)_2(Si_4O_{11})_2$ Monoclinic

$$n_\alpha = 1.657 \text{ to } 1.663$$
$$n_\beta = 1.684 \text{ to } 1.697$$
$$n_\gamma = 1.699 \text{ to } 1.717$$
$$2V = 79 \text{ to } 86°; \text{ Opt. } (-)$$
$$b = \beta \text{ or Y}, c \wedge \gamma \text{ or Z} = -10 \text{ to } -14°$$

Grunerite (not grünerite) is the name used for the minerals of the cummingtonite-grunerite series in which iron greatly predominates over magnesium.

Color.—Neutral in thin sections.

Form.—Grunerite usually occurs in fibrous to columnar aggregates. It is sometimes asbestiform. Cross sections are rhombic as in other amphiboles.

Cleavage.—Cleavage is in two directions at angles of about 56 and 124° as in the other amphiboles. Cross fractures are common.

Relief fairly high, $n >$ balsam.

Birefringence strong, $n_\gamma - n_\alpha = 0.042$ to 0.054; the interference colors range up to upper second order or low third order. Sections with parallel extinction show low first-order colors.

Extinction.—The maximum extinction angles in longitudinal sections varies from 10 up to about 15°. The variation in maximum extinction angle is due to variation in chemical composition. The replacement of some of the iron by magnesium brings about an increase in the extinction angle.

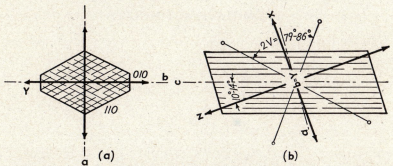

Figs. 259 *a,b.*—Orientation diagrams of grunerite. Sections (*a*) normal to the *c*-axis and (*b*) parallel to (010).

Orientation.—Elongate sections are length-slow.

Twinning.—A characteristic feature of grunerite is the polysynthetic twinning with (100) as the twin-plane. The twin-lamellae are often very narrow.

Interference Figure.—Interference figures are often difficult to obtain, but when found they are biaxial negative with a large axial angle. The axial plane is (010). Dispersion, $r > v$, weak.

Distinguishing Features.—Grunerite has a smaller maximum extinction angle and higher indices of refraction than cummingtonite. It is optically negative and cummingtonite positive. The in-

Fig. 260.—(×25) Grunerite with magnetite.

dices of refraction of grunerite are also higher than those of tremolite-actinolite. Anthophyllite, similar in many ways, has parallel extinction and never shows any twinning.

Occurrence.—Grunerite is a product of metamorphism. It occurs in metamorphic rocks such as the mica schists at Collobrières, Department Var, France (the original locality), also as alteration product in the eulysite of Tunaberg, Sweden, and is very prominent in some of the iron ores of the Lake Superior region, notably the magnetite ores of the Upper Peninsula of Michigan.

TREMOLITE-ACTINOLITE

$Ca_2(Mg,Fe)_5(OH)_2(Si_4O_{11})_2$ Monoclinic
$$\angle\beta = 74°48'$$

$$n_\alpha = 1.600 \text{ to } 1.628$$
$$n_\beta = 1.613 \text{ to } 1.644$$
$$n_\gamma = 1.625 \text{ to } 1.655$$
$$2V = 79 \text{ to } 85°; \text{ Opt. } (-)$$
$$b \doteq \beta \text{ or } Y, c \wedge \gamma \text{ or } Z = -10 \text{ to } -20°$$

Color.—Colorless to pale green in thin sections. The green varieties show faint pleochroism. Green ferriferous tremolite is known as *actinolite*.

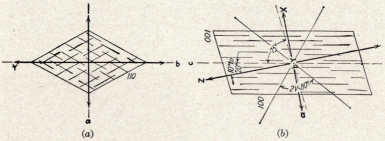

Figs. 261 *a,b.*—Orientation diagrams of tremolite-actinolite. Sections (*a*) normal to the *c*-axis and (*b*) parallel to (010).

Form.—Tremolite-actinolite occurs in long prismatic crystals and columnar to fibrous aggregates. Asbestiform varieties are common. The typical cross section is rhombic with (110 $\wedge$ 1$\bar{1}$0) = 56°.

Cleavage {110} in two directions at angles of about 56 and 124°. Longitudinal sections show cleavage traces parallel to the length. There may be parting parallel to (100).

Relief fairly high, $n >$ balsam.

Birefringence moderate to rather strong, $n_\gamma - n_\alpha = 0.022$ to 0.027; so the interference colors range up to low or middle second order. Narrow longitudinal sections show the highest colors. Cross sections have white to yellow interference colors.

Extinction.—The maximum extinction in longitudinal sections varies from 10 to 20°. A few longitudinal sections have parallel or nearly parallel extinction. Cross sections have symmetrical extinction.

Orientation.—Elongate sections are length-slow. In cross sections the long diagonal is the slower ray.

Twinning.—Twins with {100} as twin-plane are frequent. Fine polysynthetic twinning with {001} as twin-plane is occasionally encountered.

FIG. 262.—(×17) Tremolite.

Interference Figure.—Tremolite-actinolite gives a biaxial negative figure with very large axial angle. The axial plane is {010}. Dispersion, $r < v$ weak. Broad elongate sections with low interference colors give the best figure.

Distinguishing Features.—The extinction angle and amphibole cross sections are characteristic. Wollastonite has the same general appearance as tremolite, but the trace of the optic axial plane is normal to the cleavage instead of parallel to it as in tremolite.

Related Minerals.—A colorless amphibole, edenite, greatly resembles tremolite but has larger extinction angles.

Alteration.—Tremolite-actinolite is sometimes found altered to talc.

Occurrence.—Tremolite-actinolite occurs in contact-metamorphic deposits, in schists and gneisses, and in metamorphic limestones. It is also found as a replacement of pyroxene in igneous rocks.

NEPHRITE

(Jade in part)

$Ca_2(Mg,Fe)_5(OH)_2(Si_4O_{11})_2$ Monoclinic

$$n_\alpha = 1.600 \text{ to } 1.628$$
$$n_\beta = 1.613 \text{ to } 1.644$$
$$n_\gamma = 1.625 \text{ to } 1.655$$
$$2V = 79 \text{ to } 85°; \text{ Opt. } (-)$$
$$b = \beta \text{ or } Y; c \wedge \gamma \text{ or } Z = -10 \text{ to } -20°$$

Nephrite is really a tough compact variety of tremolite-actinolite, but for emphasis it is treated separately.

Color.—Colorless to gray in thin sections.

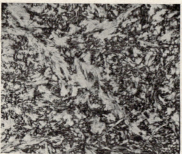

Fig. 263.—($\times 27$) Nephrite showing typical aggregate texture. ($\times$ nicols.)

Form.—Nephrite usually occurs in fibrous to fibro-lamellar aggregates of imperfect prismatic crystals.

Cleavage is like that of tremolite-actinolite but is rarely distinct on account of inter-felted fibers.

Relief fairly high, $n >$ balsam.

Birefringence moderate, $n_\gamma - n_\alpha = 0.022$ to 0.027; so the interference colors range from first-order gray up to bright colors of the middle second order.

Extinction varies from parallel to a maximum of about 10 to 20°. A few of the broader longitudinal sections may have parallel extinction. The extinction of nephrite is often wavy and indistinct.

Orientation.—Most sections are length-slow.

Twinning with {100} as twin-plane is occasionally found but does not seem to be common.

Interference Figure.—Nephrite does not usually give a good interference figure on account of the aggregate structure. The figure when obtained is biaxial negative with a large axial angle. The axial plane is {010} as in the other amphiboles.

Related Minerals.—There is no very sharp distinction between nephrite and other varieties of tremolite-actinolite. The term *seminephrite* has been used by F. J. Turner for an amphibole intermediate between nephrite and less compact, more coarsely crystalline tremolite-actinolite.

Distinguishing Features.—Nephrite is distinguished from jadeite, the other jade mineral, by its smaller maximum extinction angle and its lower indices of refraction and also by its lower specific gravity. From other varieties of tremolite-actinolite it is distinguished by its greater compactness, which is due to interfelted crystalline aggregates.

Alteration.—Nephrite is sometimes altered to talc.

Occurrence.—Nephrite usually occurs in association with serpentine as "kidneys" with more or less schistose structure and in derived water-worn pebbles and boulders. It is doubtless a product of metamorphism, but its origin is not well understood.

Nephrite is a widely distributed mineral much valued by the natives of many countries as material for both weapons and ornaments.

HORNBLENDE

$Ca_2(Mg,Fe,Al)_5(OH)_2[(Si,Al)_4O_{11}]_2$

Monoclinic
$\angle\beta = 75°2'$

$$n_\alpha = 1.614 \text{ to } 1.675$$
$$n_\beta = 1.618 \text{ to } 1.691$$
$$n_\gamma = 1.633 \text{ to } 1.701$$
$$2V = 52 \text{ to } 85°; \text{ Opt. } (-)$$
$$b = \beta \text{ or Y}, c \wedge \gamma \text{ or Z} = -12 \text{ to } -30°$$

Color.—Green or brown of various tones in thin sections. Pleochroism as follows:

α or X	β or Y	γ or Z
Yellow green	Olive green	Dark green
Pale green	Green	Dark green
Pale brown	Greenish	Dark green
Yellow green	Yellow	Brown
Greenish brown	Reddish brown	Red brown

Absorption scheme: γ or Z > β or Y > α or X.

Form.—Crystals are prismatic in habit with pseudo-hexagonal cross sections (110 ∧ 1$\overline{1}$0 = 55°49′). Crystals are rarely well terminated.

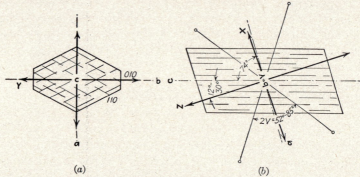

(a)

(b)

Figs. 264 *a,b.*—Orientation diagrams of hornblende. Sections (*a*) normal to the *c*-axis and (*b*) parallel to (010).

Cleavage {110} in two directions at angles of about 56 and 124°.

Relief rather high, $n >$ balsam.

Birefringence moderate, $n_\gamma - n_\alpha = 0.019$ to 0.026. The maximum interference colors are about middle second order, but in many varieties the color of the mineral modifies or even masks the interference colors.

Fig. 265.—(×13) Hornblende with plagioclase and magnetite in norite.

Extinction.—The maximum extinction angle in longitudinal sections varies from about 12 to about 30°. In cross sections the extinction is symmetrical to the outlines or to cleavage traces.

Twinning.—Twins with {100} as the twin-plane are rather common. Twinning is often manifest as twin seams.

Interference Figure.—The figure is biaxial negative with a large axial angle. The axial plane is {010}. Dispersion, $r < v$ weak.

Distinguishing Features.—Hornblende differs from augite in cleavage, pleochroism, and maximum extinction angle. Brown hornblende resembles biotite, but the latter has better cleavage (in one direction only) and parallel or almost parallel extinction. Lamprobolite has a smaller extinction angle, higher indices of refraction, and stronger birefringence.

Fig. 266.—(×12) Late magmatic hornblende (dark) formed at the expense of pyroxene.

Occurrence.—Hornblende is a very common and widely distributed mineral in many types of igneous rocks. It also occurs in schists, gneisses, and amphibolites.

It is a prominent constituent of many detrital sediments.

LAMPROBOLITE[1]

(Basaltic Hornblende)

Ca,Mg,Fe,Al silicate

Monoclinic
$\angle\beta = 73°58'$

$$n_\alpha = 1.670 \text{ to } 1.692$$
$$n_\beta = 1.683 \text{ to } 1.730$$
$$n_\gamma = 1.693 \text{ to } 1.760$$
$$2V = 64 \text{ to } 80°; \text{ Opt. } (-)$$
$$b = \beta \text{ or Y}, c \wedge \gamma \text{ or Z} = 0 \text{ to } -12°$$

Color.—Yellow to brown, often with opaque borders. Pleochroism rather strong: α or X, light yellow; β or Y, brown; γ or Z, dark red-brown.

Form.—Lamprobolite occurs almost invariably in euhedral crystals with the pseudo-hexagonal cross section of the amphiboles. The habit is usually short prismatic.

Cleavage {110} in two directions at angles of 56 and 124° as in the other amphiboles.

[1] The name *lamprobolite* was proposed by the senior author (*Am. Min.*, vol. 25, pp. 826–828, 1940) for the mineral usually called *basaltic hornblende.*

Relief high, $n >$ balsam.

Birefringence rather strong to very strong, $n_\gamma - n_\alpha = 0.026$ to 0.072. The interference colors should be high order, but they are usually masked by the color of the mineral.

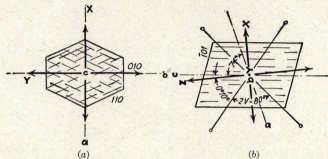

FIGS. 267 *a,b.*—Orientation diagrams of lamprobolite. Sections (*a*) normal to the *c*-axis and (*b*) parallel to (010).

Extinction.—The maximum extinction angle is very small, from zero up to as much as 12° in some varieties.

Orientation.—The crystals are length-slow. Cross sections have symmetrical extinction.

Twinning.—Twins with {100} as twin-plane are found but are not conspicuous on account of the small size of the extinction angle.

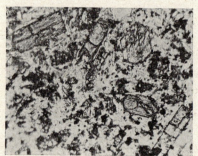

FIG. 268.—(×50) Phenocrysts of lamprobolite.

Interference Figure.—The figure is biaxial negative with a large axial angle. The axial plane is {010}. Dispersion, $r < v$.

Distinguishing Features.—Lamprobolite is distinguished from ordinary brown hornblende by the smaller extinction angle and the stronger birefringence. Biotite shows no cleavage in six-sided sections.

Related Minerals.—Kaersutite is a titanian amphibole related to lamprobolite.

Occurrence.—Lamprobolite occurs in volcanic rocks such as andesites, auganites, basalts, basanites, tephrites, and the corresponding tuffs.

It is also fairly common as a detrital mineral.

It seems likely that lamprobolite has been produced from ordinary hornblende by the oxidation of the iron, probably by hot gases at the end of the magmatic stage.

Riebeckite

(inc. Crocidolite)

$NaFe^{III}(SiO_3)_2.Fe^{II}SiO_3$

Monoclinic
$\angle\beta = 76°10'$

$$n_\alpha = 1.693$$
$$n_\beta = 1.695$$
$$n_\gamma = 1.697$$

2V large; Opt. $(-)$

$b = \beta$ or Y, $c \wedge \alpha$ or X $= +5°$

Color.—Dark blue in thin sections. Pleochroism strong: α or X, deep blue; β or Y, lighter blue; γ or Z, greenish. Absorption: α or X $> \beta$ or Y $> \gamma$ or Z.

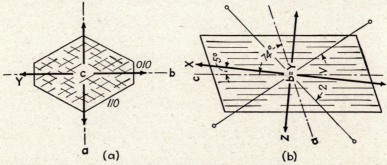

FIGS. 269 *a,b.*—Orientation diagrams of riebeckite. Sections (*a*) normal to the *c*-axis and (*b*) parallel to (010).

Form.—Riebeckite occurs in subhedral prismatic crystals and in fibrous and asbestiform aggregates. According to recent investigations, crocidolite is a fibrous variety of riebeckite.

Cleavage {110} in two directions at angles of about 56 and 124°.

Relief high, $n >$ balsam.

Birefringence very weak; $n_\gamma - n_\alpha = 0.004$; the interference colors are masked by the deep color of the mineral.

Extinction.—The maximum extinction angle in elongate sections is about 5°, but the fibrous variety, crocidolite, has parallel extinction.

Orientation.—The crystals are length-fast.

Interference Figure.—The figure is biaxial negative with a large axial angle. The axial plane is {010}. Dispersion, $r > v$ strong.

Distinguishing Features.—The color, pleochroism, and small extinction angle are distinctive.

Alteration.—Crocidolite is often altered to an iron-stained fibrous quartz known as "tiger's eye."

Occurrence.—Riebeckite is characteristic of soda-rich granites, microgranites, granite aplites, granite pegmatites, syenites, nepheline syenites, and trachytes. In these rocks it is often associated with aegirine. Crocidolite is found in certain highly siliceous metamorphic rocks such as the "ironstones" of Griqualand West, South Africa.

Glaucophane

$Na_2Mg_3Al_2(OH)_2(Si_4O_{11})_2$ Monoclinic
$\angle\beta = 77°$

$$n_\alpha = 1.621 \text{ to } 1.655$$
$$n_\beta = 1.638 \text{ to } 1.664$$
$$n_\gamma = 1.639 \text{ to } 1.668$$
$$2V = 0 \text{ to } 68°; \text{ Opt. } (-)$$
$$b = \beta \text{ or } Y; c \wedge \gamma \text{ or } Z = -4 \text{ to } -6°$$

Color.—Blue to violet in thin sections. Pleochroism: α or X, neutral; β or Y, violet; γ or Z, blue.

Form.—Glaucophane occurs in prismatic crystals or columnar aggregates. The cross sections are pseudo-hexagonal or rhombic.

Cleavage {110} in two directions at angles of 56 and 124°.

Relief fairly high, $n >$ balsam.

Birefringence moderate, $n_\gamma - n_\alpha = 0.013$ to 0.018; the maximum interference color is about sensitive violet, but the color of the mineral may modify or even mask the interference colors.

Extinction.—The maximum extinction angle in longitudinal sections is very small (4 to 6°). Cross sections have symmetrical extinction.

Orientation.—The crystals are length-slow.

Interference Figure.—The figure is biaxial negative with a small to moderate axial angle. The axial plane is {010}. Dispersion, $r < v$ strong.

Distinguishing Features.—The axial colors, together with the small extinction angle and amphibole cross section and cleavage, distinguish glaucophane from all other minerals except crossite and gastaldite.

Related Minerals.—Crossite is a soda amphibole intermediate between glaucophane and riebeckite, but the axial plane of

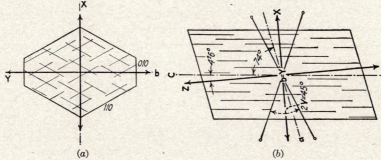

(a)   (b)

Figs. 270 *a,b.*—Orientation diagrams of glaucophane. Sections (*a*) normal to the *c*-axis and (*b*) parallel to (010).

crossite is normal to (010). Gastaldite is a soda amphibole related to glaucophane but paler blue in color because of lower iron content.

Occurrence.—Glaucophane is found in certain schists and gneisses. The usual associates are muscovite, quartz, garnet, sphene, lawsonite, and clinozoisite. Glaucophane schists are abundant in the Coast Ranges of California and are also found in Syra (Greece), Italy, and Japan, but are rare, taken the world over.

THE OLIVINE GROUP

The olivine group consists of the two end members, forsterite and fayalite, and the intermediate isomorphous mixture, olivine. The rare mineral tephroite, Mn_2SiO_4, also belongs to the olivine group. The double salt, monticellite, $(CaMgSiO_4)$, is closely related to olivine and is often considered to be a member of the group. Its crystal system is orthorhombic. Larnite, with the composition Ca_2SiO_4, is not a member of the olivine group since it is monoclinic.

THE OLIVINE GROUP

Mineral	Chemical composition	n_α	n_β	n_γ	2V
Forsterite (Fo)	Mg_2SiO_4	1.635 1.640	1.651 1.660	1.670 1.680	85–90°
Olivine (Fo$_m$Fa$_n$)	$(Mg,Fe)_2SiO_4$	1.651 1.681	1.670 1.706	1.689 1.718	70–90°
Fayalite (Fa)	Fe_2SiO_4	1.805 1.835	1.838 1.877	1.847 1.886	47–54°
Monticellite	$CaMgSiO_4$	1.641 1.651	1.646 1.662	1.655 1.669	75–80°

Minerals of the olivine group are characterized by rather high refractive indices and strong birefringence. The axial plane is (001), and the axial angle is usually very large.

Olivine is exceedingly abundant as a rock-forming mineral in subsilicic igneous rocks. Forsterite is practically limited to metamorphic limestones or contact metamorphic zones. Fayalite is found in granite pegmatites, in lithophysae of rhyolitic obsidians, and in some ores, but it is rather rare.

Forsterite

Mg_2SiO_4 Orthorhombic

$$n_\alpha = 1.635 \text{ to } 1.640$$
$$n_\beta = 1.651 \text{ to } 1.660$$
$$n_\gamma = 1.670 \text{ to } 1.680$$
$$2V = 85 \text{ to } 90°; \text{ Opt. } (+)$$
$$a = \gamma \text{ or } Z, \ b = \alpha \text{ or } X, \ c = \beta \text{ or } Y$$

Color.—Colorless in thin sections.

Form.—Forsterite usually occurs in euhedral to subhedral crystals.

Relief fairly high, $n >$ balsam. The indices increase with increasing iron content.

Cleavage {010} imperfect. Irregular fractures common.

Birefringence strong, $n_\gamma - n_\alpha = 0.035$ to 0.040. The maximum interference color is upper second order.

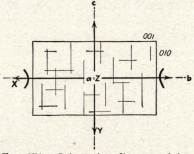

Fig. 271.—Orientation diagram of forsterite. Section parallel to (100).

Extinction parallel to crystal outlines and cleavage traces.

Orientation.—Crystals showing cleavage are length-slow.

Interference Figure.—The interference figure is biaxial positive with a very large axial angle. The axial plane is {001}. Dispersion, $r < v$.

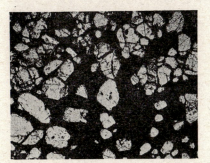

Fig. 272.—Subhedral crystals of forsterite in matrix of magnetite and ludwigite.

Distinguishing Features.—Forsterite is an iron-free olivine and so resembles ordinary olivine but has somewhat lower indices of refraction.

Alteration.—Forsterite is often altered to antigorite, but the secondary magnetite so common with altered olivine is absent.

Occurrence.—Forsterite occurs for the most part in metamorphic limestones as a product of dedolomitization. Phlogopite is a common associate. Forsterite also occurs in contact-metamorphic zones, where it is often associated with magnetite.

OLIVINE

(Mg,Fe)₂SiO₄   (Chrysolite)   Orthorhombic

$$n_\alpha = 1.651 \text{ to } 1.681$$
$$n_\beta = 1.670 \text{ to } 1.706$$
$$n_\gamma = 1.689 \text{ to } 1.718$$
$$2V = 70 \text{ to } 90°; \text{ Opt. } (+), \text{ also } (-)$$
$$a = \gamma \text{ or } Z, b = \alpha \text{ or } X, c = \beta \text{ or } Y$$

Color.—Colorless in thin sections.

Form.—Olivine occurs in anhedra with polygonal outlines and in phenocrysts with the characteristic outline of Fig. 123, page 124, which is a section parallel to {100}.

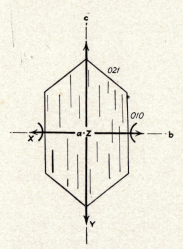

FIG. 273.—Orientation diagram of olivine. Section parallel to (100).   FIG. 274.—(×9) Twinned olivine in dunite. (× nicols.)

Cleavage imperfect parallel to {010}, irregular fractures common.

Relief fairly high, $n >$ balsam.

Birefringence strong, $n_\gamma - n_\alpha = 0.037$ to 0.041; the maximum interference color is upper second order.

Extinction parallel to crystal outlines and cleavage traces.

Orientation.—Crystals showing cleavage are length-slow.

Twinning is sometimes found, but the lamellae are broad and not well defined. The twinning is probably vicinal.

Interference Figure.—The interference figure is usually biaxial positive with a large axial angle, but olivine high in iron is optically negative. Olivine often shows a variation in the size of the axial angle within a single crystal, the angle decreasing toward the exterior of the crystal. According to Tomkeieff, this is evidence of zoning, the outer zones being richer in fayalite. The axial plane is {001}. Dispersion, $r < v$.

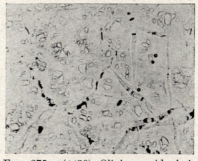

FIG. 275.—(×30) Olivine residual in serpentine with magnetite.

Distinguishing Features.—The mineral most apt to be mistaken for olivine is diopside, but diopside has better cleavage, oblique extinction, and somewhat weaker birefringence.

Related Minerals.—Olivine fairly rich in iron with about 50 per cent of Fe_2SiO_4 is known as *hyalosiderite*. The indices of refraction and the birefringence are higher than those of olivine proper.

Alteration.—Olivine commonly shows alteration to antigorite and secondary magnetite along irregular fractures (see Fig. 160, page 200). In basaltic rocks the alteration of the outer iron-rich rims of olivine to brownish-red iddingsite is fairly common (see Fig. 356, page 364).

Occurrence.—Olivine is an exceedingly common mineral in subsilicic igneous rocks such as basalts, olivine gabbros, and peridotites. In the monomineralic rock dunite it is the dominant mineral. It is a relict mineral in many serpentines (see Fig. 275).

Locally, olivine may be important as a detrital mineral.

Fayalite

$(Fe,Mg)_2SiO_4$ (Iron Olivine)   Orthorhombic

$$n_\alpha = 1.805 \text{ to } 1.835$$
$$n_\beta = 1.838 \text{ to } 1.877$$
$$n_\gamma = 1.847 \text{ to } 1.886$$
$$2V = 47 \text{ to } 54°; \text{ Opt. } (-)$$
$$a = \gamma \text{ or } Z, \, b = \alpha \text{ or } X, \, c = \beta \text{ or } Y$$

Color.—Colorless to yellowish or neutral in thin sections; it may show faint pleochroism.

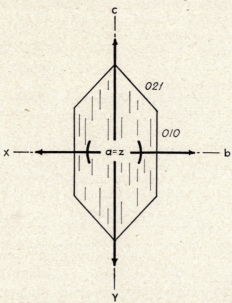

Fig. 276.—Orientation diagram of fayalite.   Section parallel to (100).

Form.—In cavities fayalite is euhedral, but as a rule it occurs in anhedral crystals.

Cleavage imperfect in one direction {010}.

Relief very high, $n >$ balsam.

Birefringence strong, $n_\gamma - n_\alpha = 0.042$ to 0.051.

Extinction parallel to cleavage traces.

Orientation.—Cleavage traces and crystals showing cleavage are length-slow.

Twinning.—Vicinal twinning with broad lamellae seems to be characteristic of fayalite (see Fig. 278) as well as of olivine.

Interference Figure.—The interference figure is biaxial negative with a moderate axial angle. The axial plane is {001} as in olivine. Dispersion, $r > v$.

Distinguishing Features.—Fayalite resembles olivine in its properties but may be distinguished by smaller axial angle, higher indices of refraction, and optically negative character.

Alteration.—Fayalite is sometimes found with grunerite as an alteration product.

Fig. 277.—(×25) Fayalite with pyroxene and garnet.

Fig. 278.—(×21) Fayalite with pyroxene and garnet. The same spot as Fig. 277. (× nicols.)

Related Minerals.—Knebelite, a manganian fayalite, and hortonolite $(Fe,Mg,Mn)_2SiO_4$ are intermediate members of the olivine group and are similar to fayalite in optical properties.

Tephroite, Mn_2SiO_4, a characteristic mineral of the Franklin Furnace and Långban ore deposits, is also similar to fayalite.

Occurrence.—Fayalite is a rather rare mineral. It is an associate of iron ores. Manganian fayalite, a low-grade iron ore in Tunaberg, Sweden, occurs in a rock called eulysite, the origin of which is in doubt. It is also found in abundance in high-temperature deposits in Santa Eulalia, Chihuahua, Mexico, according to Basil Prescott.

It is a widely distributed mineral in lithophysae of rhyolitic obsidian accompanying cristobalite.

Fayalite is one of the most characteristic constituents of furnace slags.

Monticellite

CaMgSiO$_4$ Orthorhombic

$$n_\alpha = 1.641 \text{ to } 1.651$$
$$n_\beta = 1.646 \text{ to } 1.662$$
$$n_\gamma = 1.655 \text{ to } 1.669$$
$$2V = 75 \text{ to } 80°; \text{ Opt. } (-)$$
$$a = \gamma \text{ or } Z, \ b = \alpha \text{ or } X, \ c = \beta \text{ or } Y$$

Monticellite is usually considered to be a member of the olivine group, but, strictly speaking, it is a double salt.

Color.—Colorless in thin sections.

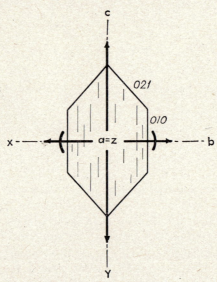

Fig. 279.—Orientation diagram of monticellite. Section parallel to (100).

Form.—Usually in granular aggregates of anhedral to subhedral crystals, but it may also occur in euhedral prismatic crystals. In some igneous rocks it is found in rims around olivine crystals.

Cleavage.—Imperfect parallel to {010}.

Relief rather high, $n >$ balsam.

Birefringence moderate, $n_\gamma - n_\alpha = 0.014$ to 0.018; so the maximum interference color is first-order red.

Extinction parallel to cleavage traces and to the main crystal outlines.

Orientation.—Crystals showing cleavage are length-slow.

Interference Figure.—The interference figure is biaxial with a large axial angle. The axial plane is (001) as in olivine. Dispersion $r > v$.

Distinguishing Features.—Monticellite is a rather difficult mineral to recognize since it has no very distinctive properties. It resembles forsterite and olivine but has weaker birefringence than either of these. It may also be distinguished from forsterite and from most olivines by its negative optical sign.

Alteration.—Some of the Crestmore monticellite is replaced by idocrase.

Occurrence.—Monticellite is a contact-metamorphic mineral usually found in limestones and dolomites. Large

Fig. 280.—($\times$11) Monticellite with idocrase (gray) in monticellite rock.

masses of monticellite constituting a veritable monticellite rock occur at Crestmore, the famous mineral locality in Riverside County, California.

Monticellite is occasionally found in igneous rocks such as alnöite (Bowen), polzenite (Scheumann), and nepheline basalt. In these rocks it occurs as overgrowths or rims in parallel position on olivine.

Chondrodite

$2Mg_2SiO_4 . Mg(OH,F)_2$ Monoclinic
$\angle \beta = 90°$

$$n_\alpha = 1.592 \text{ to } 1.643$$
$$n_\beta = 1.602 \text{ to } 1.655$$
$$n_\gamma = 1.621 \text{ to } 1.670$$
$$2V = 70 \text{ to } 90°; \text{ Opt. } (+)$$
$$b = \gamma \text{ or } Z, a \wedge \alpha \text{ or } X = -26 \text{ to } -31°$$

Color.—Colorless to yellowish or brownish. The deeper colored varieties are pleochroic from neutral to brown or pale brown to red-brown, etc.

Form.—Chondrodite is commonly found in subhedral crystals, which are often more or less rounded and in large anhedra.

Although the mineral is monoclinic, the $\angle\beta$ between the *a*- and *c*-axes is 90° (orthorhombic syngony).

Cleavage.—There is often parting parallel to (001) which is due to twinning.

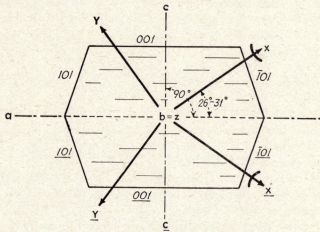

Fig. 281.—Orientation diagram of chondrodite. Twin crystal with (001) as twin-plane. Section parallel to (010).

Relief fairly high, $n >$ balsam.

Birefringence rather strong, $n_\gamma - n_\alpha = 0.027$ to 0.035; the maximum interference color varies from green to red of the second order for different varieties.

Extinction.—The maximum extinction angle measured from the trace of the twin-plane {001} varies from -26 to $-31°$.

Orientation.—The extinction direction nearest the trace of the twin-plane is the faster ray.

Twinning.—Twinning, which may appear as simple twins, twin seams, or polysynthetic twins, is rather common. The twin-plane is

Fig. 282.—(×11) Chondrodite showing polysynthetic twinning. (× nicols.)

{001}. Without twinning it is very difficult to orient chondrodite. There may also be twinning parallel to {105} and {305}.

Interference Figure.—The figure is biaxial positive with a large axial angle. The axial plane is normal to {010}. Dispersion, $r > v$ weak.

Distinguishing Features.—The pleochroism usually distinguishes chondrodite from olivine.

Related Minerals.—Three other minerals of the chondrodite group, *viz.*, norbergite, humite, and clinohumite, are similar to chondrodite. The distinction is based upon extinction angles and refractive indices.

Alleghanyite, the manganese analogue of chondrodite, is much like chondrodite, but it is optically negative.

Occurrence.—Chondrodite is one of the characteristic minerals of metamorphic limestone. It is often associated with phlogopite and spinel.

GARNET GROUP

Pyrope	$Mg_3Al_2(SiO_4)_3$	$n = 1.741$ to 1.760
Almandite	$Fe_3Al_2(SiO_4)_3$	$n = 1.778$ to 1.815
Spessartite	$Mn_3Al_2(SiO_4)_3$	$n = 1.792$ to 1.820
Uvarovite	$Ca_3Cr_2(SiO_4)_3$	$n = 1.838$ to 1.870
Grossularite	$Ca_3Al_2(SiO_4)_3$	$n = 1.736$ to 1.763
Andradite	$Ca_3Fe_2(SiO_4)_3$	$n = 1.857$ to 1.887

The six minerals of the garnet group here listed may be classified in two subgroups: the pyrope-almandite-spessartite series (called *pyralspite* by Winchell) and the uvarovite-grossularite-andradite series (called *ugrandite* by Winchell). It is rare to find a garnet that corresponds to any one of the formulae given. They are isomorphous mixtures of these end members in varying amount. The name is assigned according to the dominant end member present.

Color.—Colorless, pale reddish, pale to dark brown, greenish gray, etc., in thin sections. Crystals are often zoned.

Form.—Euhedral dodecahedral crystals in six-sided sections and trapezohedral crystals in eight-sided sections are common. Garnet also occurs in polygonal grains, aggregates, and masses. Inclusions are frequent.

Cleavage absent, but it may have parting parallel to {110}. Irregular fractures are characteristic.

Relief very high, surface rough; $n > $ balsam.

Birefringence.—Most varieties are dark between crossed nicols but some have weak or very weak birefringence. The birefringent areas are often arranged in zones or sectors (see Fig. 284).

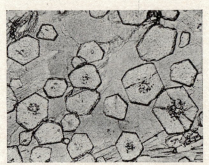

Fig. 283.—(×33) Euhedral garnet crystals in schist.

Distinguishing Features.—Garnet resembles spinel, but the latter occurs in octahedra. The different kinds of garnet may be determined by indices of refraction combined with the determination of the specific gravity.

Alteration.—The most common alteration product of garnet is chlorite.

Occurrence.—Garnet is especially characteristic of metamorphic rocks. It is also a very common detrital mineral. Almandite is the common garnet of schists and gneisses. Pyrope is practically confined to peridotites and derived serpentines. Grossularite and andradite are common in contact-metamorphic zones. Melanite, a deep brown variety of andradite, occurs in soda-rich igneous rocks such as nepheline syenites, phonolites, etc.

Fig. 284.—(×17) Zoned garnet in metamorphic limestone. (× nicols.) (*Courtesy of F. C. Gros.*)

Spessartite is found in pegmatites, schists, and quartzites.

Uvarovite, the rarest of the garnets, is found as a secondary mineral in chromite and also in some contact-metamorphic zones.

Garnet is common as a detrital mineral. The grains are often grooved or pitted.

Beryl

$Be_3Al_2(SiO_3)_6$

Hexagonal
(Hexagonal Subsystem)

$$n_\epsilon = 1.564 \text{ to } 1.590$$
$$n_\omega = 1.568 \text{ to } 1.598$$
$$\text{Opt. } (-)$$

Color.—Colorless in thin sections. In thick oriented sections colored varieties such as emerald are somewhat pleochroic.

Form.—Beryl usually occurs in rather large crystals of prismatic habit, occasionally in small, slender prisms, and also in massive form. Liquid inclusions with gas bubbles in six-sided negative crystals are common in oriented sections.

Cleavage imperfect parallel to (0001), not usually seen in thin sections.

Relief moderate, $n >$ balsam.

Birefringence weak, $n_\omega - n_\epsilon = 0.004$ to 0.008; interference colors are gray, white, or straw yellow of the first order.

Extinction.—Longitudinal sections have parallel extinction. Basal sections are dark in all positions.

Orientation.—Crystal sections are length-fast.

Interference Figure.—Basal sections give a negative uniaxial figure without any rings. The cross may show a slight opening in certain areas.

Distinguishing Features.—Beryl resembles apatite, but the latter has higher indices of refraction. From quartz it is distinguished by its length-fast character and optical sign.

Alteration.—It is sometimes altered to kaolin.

Occurrence.—The principal occurrence of beryl is in granite pegmatites. It is also found in mica schists and in veins in limestone associated with albite.

Scapolite Group

(Wernerite)

$m[3NaAlSi_3O_8.NaCl] = Ma$ Tetragonal
$n[3CaAl_2Si_2O_8.Ca(O,CO_3,SO_4)] = Me$

$$n_\epsilon = 1.540 \text{ to } 1.571$$
$$n_\omega = 1.550 \text{ to } 1.607$$
$$\text{Opt. } (-)$$

Scapolite is an isomorphous mixture of the two end members given above. The sodium end member is called *marialite*, and the calcium end member, *meionite*. The name *wernerite* is applied to certain intermediate members.

It will be noted that the chemical composition is similar to that of the plagioclases but with added $NaCl, CaCO_3$, etc.

Color.—Colorless in thin sections.

Form.—Minerals of the scapolite group usually occur in columnar aggregates. Crystals are usually rather large.

Relief low to fair, $n >$ balsam.

Cleavage distinct parallel to {100}, less distinct parallel to {110}. In most sections the cleavage traces are parallel to the length; in cross sections the cleavage shows in two directions at right angles.

Birefringence rather weak to rather strong, $n_\omega - n_\epsilon = 0.010$ to 0.036. The maximum interference color varies from yellow of the first order up to second-order violet, depending upon the chemical composition. The birefringence increases with calcium or meionite content.

Extinction parallel in most sections. Basal sections remain dark between crossed nicols.

Orientation.—The cleavage traces and main crystal outlines are parallel to the faster ray.

Interference Figure.—Basal sections give a uniaxial negative figure with a few rings. Longitudinal sections give a "flash figure."

Distinguishing Features.—Scapolite is similar to plagioclase but lacks twinning, has parallel extinction, and usually has stronger birefringence. Varieties with weaker birefringence resemble cordierite, which is biaxial, or quartz, which is optically positive. Without chemical analyses it is difficult to determine the various kinds of scapolite, but the birefringence increases with the calcium content.

Alteration.—Scapolite is often altered to muscovite and to ill-defined fibrous aggregates.

Occurrence.—The characteristic occurrence of scapolite is contact-metamorphic limestones where it is often associated with idocrase, diopside, garnet, etc. It also occurs in certain gneisses and in some gabbros as a high-temperature alteration of plagioclase.

Idocrase

$Ca_2Al_2(OH,F)Si_2O_7$ (Vesuvianite)   Tetragonal

$$n_\epsilon = 1.701 \text{ to } 1.726$$
$$n_\omega = 1.705 \text{ to } 1.732$$
$$\text{Opt. } (-)$$

Color.—Colorless to neutral in thin sections. It may be pleochroic in thick sections.

Form.—Idocrase occurs in euhedral crystals, in columnar aggregates, in anhedra with polygonal outlines, and in fine aggregates.

Cleavage imperfect parallel to {110}.

Relief high, $n >$ balsam.

Birefringence v e r y   w e a k to weak, $n_\omega - n_\epsilon = 0.004$ to 0.006; interference colors are low first-order gray, sometimes normal, sometimes anomalous gray-green, purple, or deep blue.

Fig. 285.—($\times$30) Idocrase, variety californite. ($\times$ nicols.)

Extinction parallel.

Orientation length-fast in columnar aggregates.

Interference Figure.—The figure is usually uniaxial negative but may be biaxial with a small axial angle.

Distinguishing Features.—Anomalous idocrase resembles zoisite and clinozoisite and is often difficult to distinguish from them.

Alteration rarely observed.

Occurrence.—The principal occurrence of idocrase is in contact-metamorphic zones. Associated minerals are garnet, diopside, wollastonite, epidote, and calcite. Idocrase is also found in association with serpentine as a kind of pseudojade (californite).

Zircon

ZrSiO₄   Tetragonal

$$n_\omega = 1.925 \text{ to } 1.931$$
$$n_\epsilon = 1.985 \text{ to } 1.993$$
$$\text{Opt. } (+)$$

Color.—Colorless to pale colors in thin sections.

Form.—Zircon usually occurs in minute crystals of short prismatic habit. They are often found as inclusions and may be surrounded by pleochroic haloes.

Cleavage absent.

Relief very high, $n >$ balsam.

Birefringence very strong, $n_\epsilon - n_\omega = 0.060$ to 0.062; the maximum interference colors are usually pale tints of the fourth order, but minute crystals show lower interference colors.

Extinction parallel.

Orientation.—Crystals are length-slow.

Interference Figure.—The interference figure is uniaxial but may be difficult to obtain on account of the small size of the crystals.

Distinguishing Features.—Zircon is distinguished from apatite by stronger birefringence and higher relief.

Related Minerals.—Malacon is the metamict alteration product of zircon. It is an amorphous mineraloid.

Occurrence.—Zircon is a widely distributed mineral in grained igneous rocks. In some syenites it is prominent enough to furnish the name zircon syenite. Zircon also occurs in certain metamorphic rocks, according to Gillson.

It is one of the most widespread and abundant detrital minerals. Occasionally it is a prominent mineral in sandstone.

Topaz

Al$_2$(F,OH)$_2$SiO$_4$ Orthorhombic

$$n_\alpha = 1.607 \text{ to } 1.629$$
$$n_\beta = 1.610 \text{ to } 1.631$$
$$n_\gamma = 1.617 \text{ to } 1.638$$
$$2V = 48 \text{ to } 65°; \text{ Opt. } (+)$$
$$a = \alpha \text{ or X, } b = \beta \text{ or Y, } c = \gamma \text{ or Z}$$

Color.—Colorless in thin sections.

Form.—Topaz appears in euhedral crystals of short prismatic habit, in anhedral grains, and in columnar aggregates. Negative crystals with fluid inclusions and gas bubbles are not uncommon.

Cleavage perfect in one direction parallel to {001}.

Relief fairly high, $n >$ balsam.

Birefringence rather weak, $n_\gamma - n_\alpha = 0.009$ to 0.010, about the same as that of quartz. Interference colors are gray, white, or straw yellow of the first order. Cleavage flakes show very weak birefringence since $n_\beta - n_\alpha = 0.003$.

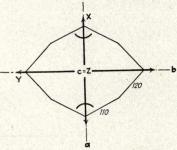

Fig. 286.—Orientation diagram of topaz. Section parallel to (001).

Extinction parallel in longitudinal sections and symmetrical in basal sections.

Orientation.—Cleavage traces are parallel to the faster ray.

Interference Figure.—Cleavage flakes and basal sections give a biaxial positive figure with a rather large axial angle. The axial plane is {010}. Dispersion, $r > v$ distinct.

Distinguishing Features.—Topaz resembles quartz but has higher relief, is biaxial, and has perfect cleavage.

Alteration to muscovite or sericite is not uncommon.

Occurrence.—Topaz occurs in high-temperature veins, in granite pegmatites, and occasionally in rhyolites. Associated minerals are tourmaline, fluorite, cassiterite (wood tin in the rhyolite occurrences), and muscovite.

Andalusite

Al$_2$SiO$_5$ (inc. Chiastolite)   Orthorhombic
(Al$_2$O$_3$.SiO$_2$)

$$n_\alpha = 1.629 \text{ to } 1.640$$
$$n_\beta = 1.633 \text{ to } 1.644$$
$$n_\gamma = 1.639 \text{ to } 1.647$$
$$2V = ca.\ 84°;\ \text{Opt. } (-)$$
$$a = \gamma \text{ or } Z,\ b = \beta \text{ or } Y,\ c = \alpha \text{ or } X$$

Color.—Usually colorless, more rarely reddish. The colored variety is pleochroic from rose-red (α or X) to pale green (β or Y) and (γ or Z).

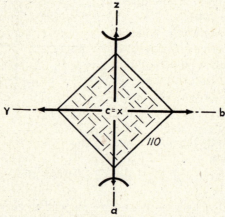

Fig. 287.—Orientation diagram of andalusite. Section parallel to (001).

Form.—Andalusite usually occurs in euhedral crystals or coarse columnar aggregates. Cross sections are nearly square (110:1$\bar{1}$0 = 89°12′). Dark inclusions of carbonaceous matter are often present and arranged symmetrically to form a kind of cross. This variety is known as *chiastolite* (see Fig. 119, page 122).

Cleavage distinct parallel to {110}. In cross sections the cleavage traces are in two directions at approximately right angles.

Relief fairly high, $n >$ balsam.

Birefringence rather weak, $n_\gamma - n_\alpha = 0.007$ to 0.011, near that of quartz. Interference colors range up to first-order yellow.

Extinction parallel in most sections. Cross sections have symmetrical extinction.

Orientation.—Crystals or columnar aggregates are length-fast.

Interference Figure.—Cross sections of crystals give a negative biaxial figure with a very large axial angle. The axial plane is {010}. Dispersion, $r > v$, weak.

Distinguishing Features.—Andalusite is distinguished from sillimanite by its length-fast character, weaker birefringence, and large axial angle. The colored pleochroic variety resembles hypersthene, but the latter is length-slow instead of length-fast.

Related Minerals.—Viridine is a manganian andalusite with higher indices of refraction and stronger birefringence than those recorded here.

Alteration.—Andalusite is often found altered to sillimanite. The variety chiastolite is usually more or less altered to sericite along the lines of included carbonaceous matter.

Occurrence.—Andalusite occurs in granite pegmatites and in high-temperature veins. In the form of chiastolite it is a characteristic contact-metamorphic mineral in schists, phyllites, and slates. It is a rather common and widely distributed mineral.

Sillimanite

Al₂SiO₅   Orthorhombic
Al_2SiO_5 Orthorhombic
($Al_2O_3.SiO_2$)

$$n_\alpha = 1.657 \text{ to } 1.661$$
$$n_\beta = 1.658 \text{ to } 1.670$$
$$n_\gamma = 1.677 \text{ to } 1.684$$
$$2V = 20 \text{ to } 30°; \text{ Opt. } (+)$$
$$a = \alpha \text{ or X}, \ b = \beta \text{ or Y}, \ c = \gamma \text{ or Z}$$

Color.—Colorless in thin sections.

Form.—Sillimanite usually occurs in small, often minute, slender prismatic crystals and in a felted mass of fibers. The crystals are often more or less bent.

The crystals are nearly square in cross section with (110 $\wedge$ 1$\bar{1}$0) = 88°15′.

Cleavage parallel to {010} but not always noticed in sections. Transverse fractures are common.

Relief fairly high, $n >$ balsam.

Birefringence moderate, $n_\gamma - n_\alpha = 0.020$ to 0.023; so the interference colors range up to second-order blue. Cross sections show very low first-order colors since $n_\beta - n_\alpha = 0.001$ to 0.009.

Extinction parallel in longitudinal sections and symmetrical in cross sections.

Orientation.—The crystals or fibers are length-slow.

Interference Figure.—On account of the small size of the crystals, good figures are rarely obtained. The axial plane is {010}. Dispersion, $r > v$ strong.

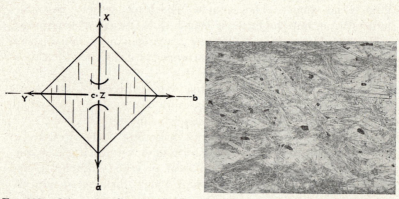

Fig. 288.—Orientation diagram of sillimanite. Section parallel to (001).

Fig. 289.—($\times$12) Sillimanite in quartz. The dark mineral is rutile.

Distinguishing Features.—Sillimanite is distinguished from andalusite by its length-slow character, stronger birefringence, and smaller axial angle. It is even more like mullite, but the latter has higher dispersion of birefringence. At times it resembles apatite, but the latter is length-fast and has weaker birefringence.

Occurrence.—Sillimanite is found in gneisses, schists, slates, hornfelses, and other metamorphic rocks. The more common associates are corundum, andalusite, kyanite, dumortierite, and cordierite.

Kyanite

Al_2SiO_5

$(Al_2O_3.SiO_2)$

Triclinic

$\angle\alpha = 90°5\frac{1}{2}'$

$\angle\beta = 101°2'$

$\angle\gamma = 105°44\frac{1}{2}'$

$n_\alpha = 1.712$

$n_\beta = 1.720$

$n_\gamma = 1.728$

$2V = ca.\ 82°$; Opt. $(-)$

Ax. pl. almost $\perp$ {100}; $c \wedge \gamma$ or $Z -30°\pm$

Color.—Colorless to pale blue. It may be pleochroic in thick sections.

Form.—The characteristic sections of kyanite are broad elongate plates tabular parallel to (100) and narrow sections parallel to (010). Crystals are often bent.

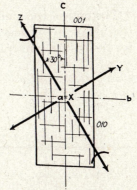

Cleavage perfect parallel to {100}, less perfect parallel to {010}, also cross parting {001} at angles of 85° with the length of the crystals.

Relief high, $n >$ balsam.

Birefringence moderate, $n_\gamma - n_\alpha = 0.016$; hence interference colors range up to first-order red.

Extinction angle on {100} is about 30° with the length of the crystals. In other sections parallel to the c-axis the extinction angle is small, sometimes practically zero. In cross sections the extinction is parallel or almost parallel.

Fig. 290.—Orientation diagram of kyanite. Section parallel to (100).

Orientation.—The extinction direction nearest the c-axis is the slow ray.

Twinning.—Twinning is frequent; there are two common twin-laws: (1) {100} = twin-plane, (2) {001} = twin-plane.

Interference Figure.—Sections cut parallel to {100} or cleavage flakes give a negative biaxial figure with a large axial angle. The axial plane makes an angle of 30° with the trace of the c-axis. Dispersion, $r > v$, weak.

Distinguishing Features.—The extinction angle of 30° together with the biaxial interference figure obtained from the broad sections is distinctive for kyanite.

Occurrence.—Kyanite occurs in schists and gneisses associated with quartz, muscovite, garnet, staurolite, and rutile. It never occurs in igneous rocks. It is also found as a detrital mineral.

Mullite

$3Al_2O_3.2SiO_2$ Orthorhombic

$$n_\alpha = 1.642$$
$$n_\beta = 1.644$$
$$n_\gamma = 1.654$$
$$2V = 20°; \text{ Opt. } (+)$$
$$a = \alpha \text{ or } X, b = \beta \text{ or } Y, c = \gamma \text{ or } Z$$

The optical constants here given are for the artificial mineral; the indices, birefringence, and optic angle are all a little higher for the natural mineral.

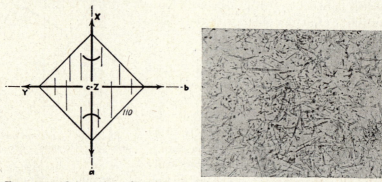

Fig. 291.—Orientation diagram of mullite. Section parallel to (001).   Fig. 292.—(×57) Mullite in feldspar of a xenolith.

Color.—Colorless in thin sections.

Form.—Mullite occurs in crystals of long prismatic habit with nearly square cross section. $110 \wedge 1\bar{1}0 = 89°13'$.

Cleavage {010}, distinct.

Relief rather high, $n >$ balsam.

Birefringence rather weak, $n_\gamma - n_\alpha = 0.012$; the maximum interference color is about first-order yellow. Cross sections have dark gray interference colors since $n_\beta - n_\alpha = 0.002$.

Extinction parallel in longitudinal sections and symmetrical in cross sections.

Orientation.—The crystals are length-slow.

Interference Figure.—The figure is biaxial positive with a moderate axial angle. The axial plane is {010}. Dispersion, $r > v$.

Distinguishing Features.—Mullite is so much like sillimanite in its properties that it was not recognized as a distinct mineral until about two decades ago. The refractive indices of sillimanite are a little higher than those of mullite.

Fig. 293.—(×16) Artificial mullite.

Occurrence.—Mullite occurs in fused argillaceous sediments found as inclusions (xenoliths) in igneous intrusions. Mullite is a very rare mineral found on the island of Mull off the west coast of Scotland.

Artificial Mullite

Artificial mullite is the substance formed by heating sillimanite, andalusite, or kyanite to a high temperature. It is used in the manufacture of high-grade porcelains such as those used in spark plugs for automobiles. Mullite is also found in ordinary porcelain as minute prismatic crystals.

Dumortierite

$HBAl_8Si_3O_{20}$ Orthorhombic

$$n_\alpha = 1.659 \text{ to } 1.678$$
$$n_\beta = 1.684 \text{ to } 1.691$$
$$n_\gamma = 1.686 \text{ to } 1.692$$
$$2V = 20 \text{ to } 40°; \text{ Opt. } (-)$$
$$a = \gamma \text{ or } Z, \, b = \beta \text{ or } Y, \, c = \alpha \text{ or } X$$

Color.—Colorless to blue, lavender, pink, or reddish. Pleochroic from colorless to blue or colorless to reddish with the greatest absorption when the length of the crystal is parallel to the vibration plane of the lower nicol. Minute crystals may not show pleochroism.

Form.—Dumortierite occurs in prismatic to acicular crystals, which often form a felt of fibers. Cross sections are pseudo-hexagonal on account of twinning (see center of Fig. 295).

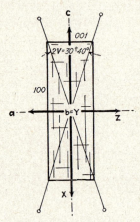

Fig. 294.—Orientation diagram of dumortierite. Section parallel to (010).

Fig. 295.—(×25) Dumortierite with quartz and rutile.

Cleavage imperfect parallel to the length. There are also cross fractures.

Relief high, $n >$ balsam.

Birefringence rather weak to moderate, $n_\gamma - n_\alpha = 0.011$ to 0.020; so the maximum inter-ference color varies from orange of the first order up to blue of the second order.

Fig. 296.—(×12) Dumortierite in schist.

Extinction parallel in most sections.

Orientation.—The crystals are length-fast.

Twinning.—Cross sections are sometimes penetration trillings with {110} as the twin-plane.

Interference Figure.—The interference figure is biaxial nega-tive with a moderate axial angle, but on account of the small size of the crystals it may be difficult to obtain. The axial plane is {010}. Dispersion, $r < v$ or $r > v$.

Distinguishing Features.—Dumortierite resembles some varieties of tourmaline, but the greater absorption is manifest when the crystals are parallel to the vibration plane of the lower nicol instead of normal to that direction. Non-pleochroic dumortierite resembles sillimanite, but the latter is length-slow.

Alteration.—Dumortierite is sometimes more or less altered to sericite.

Occurrence.—Dumortierite occurs in granite pegmatites, schists, gneisses, and other metamorphic rocks. The common associates are quartz, muscovite, tourmaline, andalusite, sillimanite, topaz, and rutile.

TOURMALINE GROUP

Nearly all minerals of the tourmaline group may be assigned to one of three divisions: (1) *schorlite* or iron tourmaline, which is black or nearly so, (2) *dravite* or magnesium tourmaline, which is brown, and (3) *elbaite* or alkali tourmaline, which is pink, red, blue, or green.

In thin sections it is not always possible to determine which tourmaline is under observation, but in the majority of cases the determination may be made.

SCHORLITE

(Iron Tourmaline)

$NaFe_3B_3Al_3(OH)_4(Al_3Si_6O_{27})$
Hexagonal
(Rhombohedral Subsystem)

$$n_\epsilon = 1.628 \text{ to } 1.658$$
$$n_\omega = 1.652 \text{ to } 1.698$$
$$\text{Opt. } (-)$$

Color.—In thin sections schorlite is variable in color: neutral gray, slate blue, buff, olive, etc. Pleochroism is usually marked, $\omega > \epsilon$. Zonal structure is rather common, especially in cross sections.

Form.—Schorlite occurs in short to long prismatic crystals and in columnar to fibrous aggregates that are more or less radiating. Spherulitic aggregates known as *tourmaline suns* are characteristic of luxullianite, a variety of tourmalinized granite from Cornwall. Cross sections of euhedral crystals are usually three sided and somewhat curved like a spherical triangle, but hexagonal cross sections are not uncommon. The polar character of the *c*-axis of crystals is rarely apparent in rock sections.

Cleavage absent, but cross fractures are very common.   There may be parting parallel to the length of the crystals.

Relief high, $n >$ balsam.

Birefringence moderate to strong, $n_\omega - n_\epsilon = 0.022$ to 0.040, usually about 0.025. The maximum interference color ranges from lower second order to lower third order, but the interference colors are more or less masked by the color of the mineral.

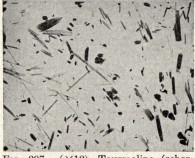

Extinction parallel in most sections. Cross sections are dark and remain dark in all positions.

Orientation.—The crystals are practically always length-fast, but it is not always possible to make this test on account of the masking of the interference colors.

Fig. 297.—(×12) Tourmaline (schorlite) in quartz.

Interference Figure.—Basal sections of the usual thickness give a negative uniaxial figure with one or two rings.   Occasionally basal sections transmit so little light that they fail to give a figure.

Distinguishing Features.—Of the more common minerals, schorlite resembles biotite and hornblende. Both of these have perfect cleavage, which is lacking in tourmaline. From these and nearly all other minerals, schorlite may be distinguished by the fact that it has the greatest absorption when the length of the crystals is normal to the vibration plane of the polarizer.

From other members of the tourmaline group, schorlite is distinguished by higher refractive indices and by its stronger absorption and pleochroism, which is lacking in thin sections of elbaite.

Occurrence.—Schorlite reaches its maximum development in granite pegmatites.   It is also found in tourmalinized granites, in greisen, and in high-temperature veins, where it is often associated with cassiterite.   It is also characteristic of certain schists and gneisses and is frequently found in hornfelses.

Schorlite is a widely distributed detrital mineral, both in fragmentary prisms and well-rounded grains.

Dravite

(Magnesium Tourmaline)

$NaMg_3B_3Al_3(OH)_4(Al_3Si_6O_{27})$ Hexagonal
(Rhombohedral Subsystem)

$$n_\epsilon = 1.613 \text{ to } 1.628$$
$$n_\omega = 1.632 \text{ to } 1.655$$
$$\text{Opt. } (-)$$

Color.—Dravite is colorless to pale yellow in thin sections and somewhat pleochroic with absorption: $\omega > \epsilon$.

Form.—Dravite usually occurs in euhedral crystals, which are generally large.

Cleavage absent, but irregular fractures are common.

Relief fairly high, $n >$ balsam.

Birefringence moderate, $n_\omega - n_\epsilon = 0.019$ to 0.025; so the maximum interference color varies from about the first sensitive violet up to a middle second-order color.

Extinction parallel to the prism faces. Occasional cross sections are dark in all positions.

Orientation.—Crystals are length-fast.

Interference Figure.—Basal sections give a negative uniaxial figure with one ring.

Distinguishing Features.—Dravite is distinguished from garnet by its moderate birefringence and pleochroism. It is distinguished from schorlite by its lower indices of refraction.

Alteration.—Dravite is sometimes found altered to an aggregate of muscovite and biotite.

Occurrence.—Dravite usually occurs in metamorphic limestones. It is also found in some schists.

Elbaite

(Alkali Tourmaline)

$Na_2Li_3B_6Al_9(OH)_8(Al_3Si_6O_{27})_2$ Hexagonal
                                             (Rhombohedral Subsystem)

$$n_\epsilon = 1.615 \text{ to } 1.629$$
$$n_\omega = 1.635 \text{ to } 1.655$$
$$\text{Opt. } (-)$$

Color.—Elbaite is colorless in thin sections. Thick sections may show pleochroism: ϵ, colorless; ω, pink, pale green, or pale blue.

Form.—Elbaite usually occurs in prismatic crystals with triangular or hexagonal cross sections.

Cleavage absent, but cross fractures are common.

Relief high, $n >$ balsam.

Birefringence moderate, $n_\omega - n_\epsilon = 0.015$ to 0.023; so the maximum interference colors vary from upper first order to lower second order. Cross sections show no birefringence.

Extinction parallel in most sections. Cross sections are dark and remain dark.

Orientation.—Elongate sections are length-fast since the habit of the crystals is prismatic and the optic sign negative.

Interference Figure.—The interference figure is uniaxial negative, usually with one ring.

Distinguishing Features.—Elbaite differs from schorlite in lighter color, weaker absorption and pleochroism, and lower refractive indices.

Occurrence.—The only important occurrence of elbaite is in granite pegmatites. The usual associates are lepidolite, albite, and quartz.

Axinite

H(Fe,Mn)Ca$_2$Al$_2$B(SiO$_4$)$_4$ Triclinic

$$n_\alpha = 1.678 \text{ to } 1.684$$
$$n_\beta = 1.685 \text{ to } 1.692$$
$$n_\gamma = 1.688 \text{ to } 1.696$$
$$2V = 70 \text{ to } 75°; \text{ Opt. } (-)$$
$$\alpha \text{ or } X \text{ almost } \perp \text{ to } (011)$$

Color.—Colorless to pale violet in thin sections.   It may show pleochroism in thick sections.

Form.—Axinite usually occurs in anhedral crystals with acute-angled sections.   Inclusions are frequent.

Cleavage imperfect in several directions.

Relief fairly high, $n >$ balsam.

Birefringence rather weak, $n_\gamma - n_\alpha = 0.010$ to 0.012, a little higher than that of quartz.

Extinction oblique to out-lines and to cleavage traces.

Fig. 298.—($\times$11) Axinite (high relief) with quartz.

Interference Figure.—The figure is biaxial negative with a large axial angle.   Dispersion, $r < v$ or $r > v$.

Distinguishing Features.—Axinite has no very distinctive features and is rather difficult to recognize in thin sections.   Its birefringence is like that of quartz, but its refractive indices are considerably higher.   It is biaxial, whereas quartz is uniaxial.

Occurrence.—Axinite occurs in the calcareous rocks of contact-metamorphic zones.   The more common associates are in addition to quartz and calcite, garnet and hedenbergite.   Axinite is also found in granites and granite pegmatites.   It is a comparatively rare mineral, but in a contact-metamorphic rock known as limurite it forms more than 50 per cent of the rock.

Zoisite

$Ca_2(Al,Fe)_3(OH)(SiO_4)_3$ Orthorhombic

$$n_\alpha = 1.696 \text{ to } 1.700$$
$$n_\beta = 1.696 \text{ to } 1.703$$
$$n_\gamma = 1.702 \text{ to } 1.718$$
$$2V = 30 \text{ to } 60°; \text{ Opt. } (+)$$

Two orientations:

(1) $a = \gamma$ or Z, $b = \beta$ or Y, $c = \alpha$ or X
(2) $a = \gamma$ or Z, $b = \alpha$ or X, $c = \beta$ or Y

There are two varieties of zoisite: a non-ferrian variety with orientation (1) and anomalous interference colors, and a ferrian variety with orientation (2) and normal interference colors.

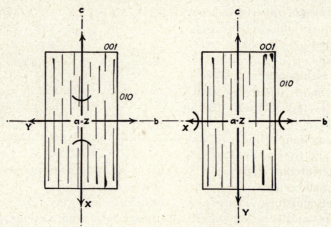

Fig. 299.   Fig. 300.

Figs. 299, 300.—Orientation diagrams of zoisite. Sections parallel to (100). Fig. 299—Orientation (1). Fig. 300—Orientation (2).

Color.—Usually colorless in thin sections, but manganian zoisite (thulite) is pink and pleochroic.

Form.—Zoisite usually occurs in columnar aggregates, but euhedral crystals are not uncommon.

Cleavage perfect in one direction {010}.

Relief high, $n >$ balsam.

Birefringence weak to moderate, $n_\gamma - n_\alpha = 0.006$ to 0.018; the interference colors in one variety (2) are normal; in the other variety (1) they are anomalous (deep blue).

Extinction parallel in most sections.

Orientation.—In some specimens (1) the crystals are length-fast; in others (2) either length-fast or length-slow.

Twinning.—Polysynthetic twinning may be present.

Interference Figure.—The interference figure is biaxial positive with a moderate axial angle. The axial plane is either (1) {010} or (2) {001}. Dispersion, (1) $r < v$ distinct or (2) $r > v$ distinct.

Distinguishing Features.— Ferrian zoisite (orientation 2) is distinguished from clinozoisite by normal interference colors. Non-ferrian zoisite (orientation 1) is distinguished from clinozoisite by a smaller axial angle and by deep blue anomalous interference color.

Fig. 301.—(×12) Euhedral zoisite crystals in quartz.

Occurrence.—Zoisite is a rather rare mineral found in some metamorphic rocks. Clinozoisite is much more common than zoisite.

Clinozoisite

$Ca_2Al_3(OH)(SiO_4)_3$ (Iron-free Epidote)   Monoclinic
$\angle\beta = 64°30'$

$$n_\alpha = 1.710 \text{ to } 1.723$$
$$n_\beta = 1.715 \text{ to } 1.729$$
$$n_\gamma = 1.719 \text{ to } 1.734$$
$$2V = 66 \text{ to } 90°; \text{ Opt. } (+)$$
$$b = \beta \text{ or } Y, c \wedge \alpha \text{ or } X = 0 \text{ to } +12°$$

Color.—Colorless and non-pleochroic in thin sections.

Form.—Clinozoisite, an iron-free or iron-poor epidote, usually occurs in elongated crystals or columnar aggregates. Cross sections are six sided, with $(100 \wedge 001) = 64\frac{1}{2}°$.

Cleavage perfect in one direction {001}.

Relief high, $n >$ balsam.

Birefringence weak to rather weak, $n_\gamma - n_\alpha = 0.005$ to 0.011; interference colors are middle first order but anomalous. The gray is somewhat blue, white is absent, and the yellow is greenish yellow.   Upper first-order colors of thicker sections are normal.

Extinction.—In most sections the extinction is parallel since the crystals are nearly always elongated in the direction of the *b*-axis.

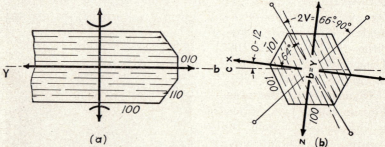

FIGS. 302 *a,b.*—Orientation diagrams of clinozoisite.   Sections (*a*) normal to the *c*-axis and (*b*) parallel to (010).

Orientation.—Some sections are length-slow and some length-fast since $b = \beta$ or Y.

Twinning.—Polysynthetic twinning with {100} as twin-plane may be found in some specimens.

Interference Figure.—The interference figure is biaxial positive with large to very large axial angle.   The axial plane is {010}.   Dispersion, $r < v$ strong.

FIG.   303.—(×15)   Clinozoisite   with glaucophane in schist.

Distinguishing Features.—Clinozoisite is distinguished from epidote by weaker birefringence, lack of pleochroism, and optically positive sign; from zoisite by the distinctive yellow-green interference color of the first order and larger axial angle.

Occurrence.—The occurrence of clinozoisite is practically the same as that of epidote.   It is a rather common and widely distributed mineral and has often been identified as zoisite.   It is usually a deuteric mineral in igneous rocks.   Clinozoisite is also found as a product of dynamic metamorphism in such rocks as amphibolites, hornblende schists, and glaucophane schists.

EPIDOTE

(Pistacite)

$Ca_2(Al,Fe)_3(OH)(SiO_4)_3$

Monoclinic
$\angle\beta = 64°37'$

$$n_\alpha = 1.720 \text{ to } 1.734$$
$$n_\beta = 1.724 \text{ to } 1.763$$
$$n_\gamma = 1.734 \text{ to } 1.779$$
$$2V = 69 \text{ to } 89°; \text{ Opt. } (-)$$
$$b = \beta \text{ or } Y, c \wedge \alpha \text{ or } X = +1 \text{ to } +5°$$

Color.—In thin sections it is colorless to yellowish green, not usually uniform. The mineral is somewhat pleochroic.

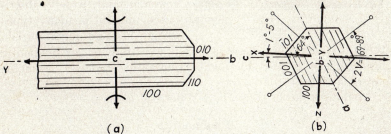

Figs. 304 *a,b*.—Orientation diagrams of epidote. Sections (*a*) normal to the *c*-axis and (*b*) parallel to (010).

Form.—Epidote occurs in granular to columnar aggregates and in more or less distinct crystals that are elongated in the direction of the *b*-axis and have a pseudo-hexagonal cross section with the forms {001}, {100}, and {$\bar{1}$01}. (100 $\wedge$ 001) = 64°37'.

Cleavage perfect in one direction {001}.

Relief high, $n >$ balsam.

Birefringence moderate to strong, $n_\gamma - n_\alpha = 0.014$ to 0.045, increasing with increase in iron content. The maximum interference colors range from low second-order to upper third-order colors. The middle first-order colors are anomalous like those of clinozoisite.

Extinction parallel in elongate sections since epidote, unlike most monoclinic crystals, is elongated in the direction of the *b*-axis.

Orientation.—Since $b = \beta$ or Y, some longitudinal sections are length-slow and some length-fast.

Twinning.—Twins with {100} as twin-plane are not uncommon.

Interference Figure.—The interference figure is biaxial negative with a large axial angle. Cleavage flakes give an optic-axis figure since one of the optic axes is almost normal to {001}.

FIG. 305.—(×60) Epidote showing euhedral crystals in epidosite.

The axial plane is {010}. Dispersion, $r > v$.

Distinguishing Features.—Epidote is distinguished from clinozoisite and zoisite by stronger birefringence and from diopside and augite by parallel extinction.

Occurrence.—Epidote is a common and widely distributed mineral in many types of igneous and metamorphic rocks. In igneous rocks it is usually a deuteric or late magmatic mineral. It is the dominant mineral in epidosite, a metamorphic epidote-quartz rock.

Epidote is rather common as a detrital mineral.

Piedmontite

(Manganese Epidote)

$Ca_2(Al,Fe,Mn)_3(OH)(SiO_4)_3$ Monoclinic
$$\angle\beta = 64°39'$$

$$n_\alpha = 1.745 \text{ to } 1.758$$
$$n_\beta = 1.764 \text{ to } 1.789$$
$$n_\gamma = 1.806 \text{ to } 1.832$$
$$2V = 56 \text{ to } 86°; \text{ Opt. } (+)$$
$$b = \beta \text{ or } Y, c \wedge \alpha \text{ or } X = -5 \text{ to } -7°$$

Color.—Vivid characteristic axial colors: yellow, orange, red, violet. Pleochroic: α or X, yellow to orange; β or Y, amethyst to violet; γ or Z, carmine to deep red.

Form.—In form piedmontite is very much like epidote.

Cleavage in one direction {001}.

Relief high, $n >$ balsam.

Birefringence very strong, $n_\gamma - n_\alpha = 0.061$ to 0.082. The interference colors are high order but are more or less masked by the color of the mineral.

Extinction parallel in elongate sections since the crystals, like those of epidote, are elongated in the direction of the *b*-axis.

Orientation.—The direction of the faster or slower ray is difficult to determine.

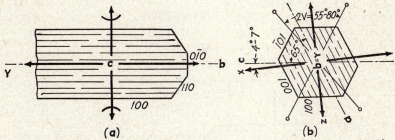

Figs. 306 *a,b.*—Orientation diagram of piedmontite. Sections (*a*) normal to the *c*-axis and (*b*) parallel to (010).

Interference Figure.—The figure is biaxial positive with a large axial angle. The axial plane is {010}. Dispersion, $r > v$ strong.

Fig. 307.—(×20) Piedmontite in quartz.

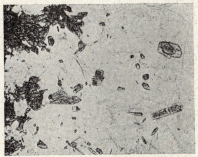

Fig. 308.—(×26) Piedmontite with overgrowth of clinozoisite in a matrix of quartz.

Distinguishing Features.—The color and pleochroism of piedmontite are so distinctive that there is little chance of mistaking it for any other mineral.

Occurrence.—Piedmontite occurs for the most part in schists and gneisses, also in altered quartz porphyries, as at South Mountain, Pa.

Allanite

(Orthite)

$$(Ca,Fe^{II})_2(Al,Ce,Fe^{III})_3(OH)(SiO_4)_3$$

Monoclinic
$\angle\beta = 65°$

$$n_\alpha = 1.64 \text{ to } 1.77$$
$$n_\beta = 1.65 \text{ to } 1.77$$
$$n_\gamma = 1.66 \text{ to } 1.80$$
$$2V = \text{large; Opt. } (-)$$
$$b = \beta \text{ or } Y, c \wedge \alpha \text{ or } X = +36°$$

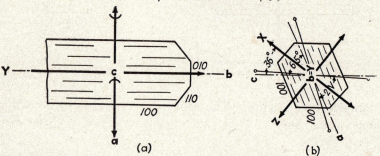

Figs. 309 *a,b.*—Orientation diagrams of allanite. Sections (*a*) normal to the *c*-axis and (*b*) parallel to (010).

Color.—Brown and pleochroic from pale brown to dark brown in thin sections.

Form.—In form allanite is similar to epidote, of which it is a cerium-bearing variety. It often occurs in parallel position as an overgrowth on epidote.

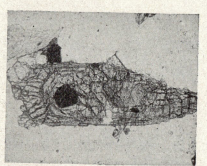

Fig. 310.—(×20) Allanite in granite.

Relief high, $n >$ balsam.

Cleavage imperfect parallel to {001}.

Birefringence rather strong, $n_\gamma - n_\alpha = 0.01$ to 0.03; the interference colors are usually masked by the brown color of the mineral.

Extinction usually parallel, like other members of the epidote group.

Orientation difficult to obtain.

Twinning is like that of epidote.

Distinguishing Features.—Allanite is distinguished from brown hornblende by parallel extinction and cleavage in one direction instead of two.

Related Minerals.—Magnesium orthite is a rare magnesian variety of allanite.

Alteration.—Allanite is often altered or inverted to an amorphous substance with about the same chemical composition as allanite. This metamict mineraloid is produced by the breakdown of the space lattice by radioactive emanations.

Occurrence.—Allanite is found in granites, syenites, granite pegmatites, and gneisses. _____

Staurolite

$2Al_2SiO_5.Fe(OH)_2$ Orthorhombic

$$n_\alpha = 1.736 \text{ to } 1.747$$
$$n_\beta = 1.741 \text{ to } 1.754$$
$$n_\gamma = 1.746 \text{ to } 1.762$$
$$2V = 80 \text{ to } 88°; \text{ Opt. } (+)$$
$$a = \beta \text{ or } Y, \; b = \alpha \text{ or } X, \; c = \gamma \text{ or } Z$$

Color.—Pale yellow in thin sections. Pleochroism distinct from nearly colorless to yellow-brown. Absorption: γ or $Z > \beta$ or $Y > \alpha$ or X.

Form.—Staurolite usually occurs in euhedral crystals of short prismatic habit and six-sided cross section with the forms $\{110\}$ and $\{010\}$.

$$(110 \wedge 1\bar{1}0) = 51°.$$

The crystals are usually a centimeter or more long.

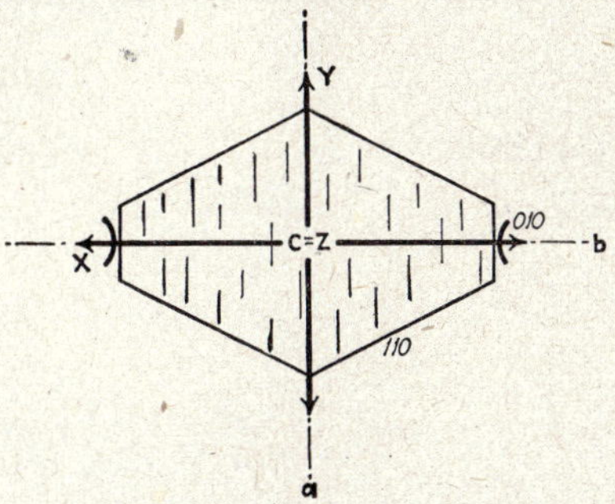

Relief high, $n >$ balsam.

Cleavage inconspicuous parallel to (010).

Fig. 311.—Orientation diagram of staurolite. Section parallel to (001).

Inclusions.—Irregularly arranged inclusions of quartz are nearly always prominent.

Birefringence rather weak, $n_\gamma - n_\alpha = 0.010$ to 0.015; the maximum interference color is first-order yellow to red.

Extinction parallel in most sections, symmetrical in cross sections.

Orientation.—The crystals are length-slow.

Twinning.—Penetration twins with {023} or {232} as twin-planes are common, but polysynthetic twins are unknown. Twinning is rarely noted in thin sections.

Interference Figure.—The interference figure is biaxial positive with a very large axial angle. The axial plane is {100}. Dispersion, $r > v$ weak.

Fig. 312.—(×16) Staurolite showing zonal structure and inclusions of quartz.

Distinguishing Features.—The color, pleochroism, and quartz inclusions are distinctive.

Occurrence.—Staurolite is found as metacrysts in metamorphic rocks such as schists, phyllites, and gneisses. Common associates are garnet, kyanite, and sillimanite in addition to quartz.

The presence of staurolite proves that the original rock was a sedimentary one.

It is also a common detrital mineral.

Sphene

CaTiSiO$_5$ (Titanite)   Monoclinic
$\angle \beta = 60°17'$

$$n_\alpha = 1.887 \text{ to } 1.913$$
$$n_\beta = 1.894 \text{ to } 1.921$$
$$n_\gamma = 1.979 \text{ to } 2.054$$
$$2V = 23 \text{ to } 50°; \text{ Opt. } (+)$$
$$b = \beta \text{ or } Y, c \wedge \alpha \text{ or } X = +33 \text{ to } +43°$$

Color.—Almost colorless to neutral in thin sections. Some varieties are pleochroic in thick sections. Axial colors: α or X, nearly colorless; β or Y, pale yellow to pale greenish; γ or Z, yellow to red-brown.

Form.—Sphene usually occurs in euhedral crystals that have an acute rhombic cross section or in irregular grains.

Cleavage.—Sphene often has prominent parting (parallel to 221). These parting directions are not parallel to the crystal outlines.

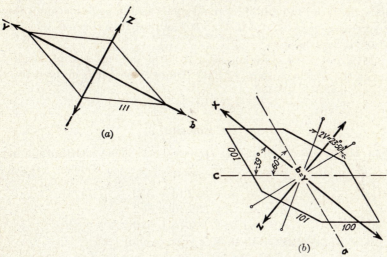

FIGS. 313 *a,b*.—Orientation diagrams of sphene.

Relief very high, $n >$ balsam.

Birefringence extreme, $n_\gamma - n_\alpha = 0.092$ to 0.141; the interference colors are high-order white but are usually obscured by total reflection.

Extinction.—On account of strong dispersion sphene does not always show complete extinction. Rhombic sections have symmetrical extinction.

Twinning.—Twins with {100} as twin-plane are sometimes present. Polysynthetic twinning parallel to (221) may also be present.

Interference Figure.—The figure is biaxial positive with a moderate axial angle. The axial plane is {010}. The acute bisectrix is almost normal to 102. Dispersion, $r > v$ strong.

FIG. 314.—($\times 25$) Euhedral crystal of sphene in granite.

Distinguishing Features.—Monazite is somewhat like sphene but has lower birefringence and weaker dispersion. The acute rhombic cross sections of sphene are very characteristic.

Occurrence.—Sphene is a widely distributed accessory (probably deuteric) mineral in grained rocks and in such metamorphic rocks as gneisses and schists. It has probably formed at a late stage in igneous rocks.

It is not very common as a detrital mineral except locally, as on the southern shore of Lake Tahoe, where it is derived from granodiorites.

Cordierite

$Mg_2Al_4Si_5O_{18}$	(Iolite)	Orthorhombic
		(Pseudo-hexagonal)

$$n_\alpha = 1.532 \text{ to } 1.552$$
$$n_\beta = 1.536 \text{ to } 1.562$$
$$n_\gamma = 1.539 \text{ to } 1.570$$
$$2V = 40 \text{ to } 80°; \text{ Opt. } (-) \text{ or } (+)$$
$$a = \beta \text{ or Y, } b = \gamma \text{ or Z, } c = \alpha \text{ or X}$$

Color.—Colorless in thin sections. Very thick sections are pleochroic. α or X, yellow; β or Y, dark violet or blue; γ or Z, pale blue or violet. Absorption: β or Y $>$ γ or Z $>$ α or X.

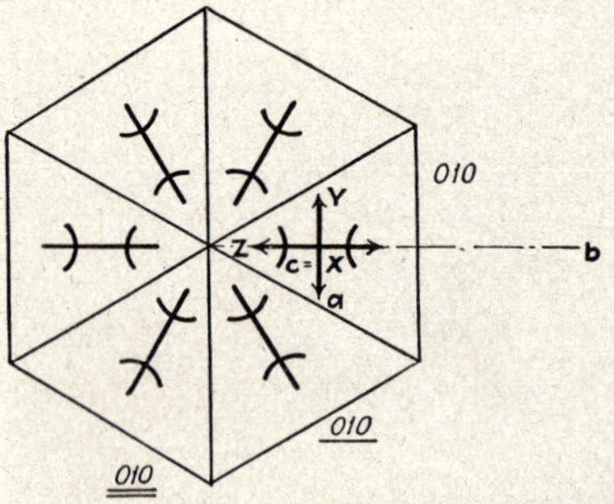

Form.—The characteristic form of cordierite is in pseudo-hexagonal crystals of short prismatic habit. These crystals are penetration twins. Cordierite also occurs in anhedra and anhedral aggregates.

Inclusions are common; these are often surrounded by pleochroic haloes.

Fig. 315.—Orientation diagram of cordierite. Section of twin-crystal parallel to (001). The twin-plane is {110}.

Cleavage imperfect parallel to (010), but it may not show in sections. Parting parallel to {001} that is due to alteration.

Relief low, n either a little less or a little greater than balsam.

Birefringence rather weak, $n_\gamma - n_\alpha = 0.007$ to 0.011, about the same as that of quartz; hence maximum interference colors are usually about straw yellow of the first order.

Extinction parallel to crystal outlines.

Twinning.—The pseudo-hexagonal crystals are penetration twins with {110} as twin-plane. Twin-lamellae are also often present.

Interference Figure.—The figure is usually biaxial negative with a variable axial angle. Optically positive cordierite has been described from several localities. The axial plane is {100}. Dispersion, $r < v$ weak.

Fig. 316.—(×30) Cordierite with polysynthetic twinning.

Distinguishing Features.—Cordierite is one of the few minerals that is easily mistaken for quartz. It is biaxial with a moderate to large axial angle and often shows twinning (either penetration or polysynthetic).

Alteration.—Cordierite is usually more or less altered to sericite (pinite), chlorite, talc, or indefinite silicates.

Occurrence.—Cordierite is a typical metamorphic mineral. It is found in gneisses and schists, often at the contact with persilicic igneous rocks. Sillimanite is a common associate. It is a characteristic mineral of hornfels. Rarely is it found in igneous rocks as an endomorphic mineral.

Prehnite

$H_2Ca_2Al_2(SiO_4)_3$ Orthorhombic

$$n_\alpha = 1.615 \text{ to } 1.635$$
$$n_\beta = 1.624 \text{ to } 1.642$$
$$n_\gamma = 1.645 \text{ to } 1.665$$
$$2V \text{ variable; Opt. } (+)$$
$$a = \alpha \text{ or } X, \ b = \beta \text{ or } Y, \ c = \gamma \text{ or } Z$$

Color.—Colorless in thin sections.

Form.—Prehnite usually occurs in aggregates that are often sheaf-like and approach spherulites. What may be called *bow-tie* structure is characteristic (Fig. 318). Crystals are mostly tabular parallel to {001}.

Cleavage good in one direction {001}.

Relief fairly high, $n >$ balsam.

Birefringence moderate to rather strong, $n_\gamma - n_\alpha = 0.020$ to 0.033; the maximum interference color varies from low

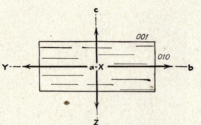

FIG. 317.—Orientation diagram of prehnite. Section parallel to (100).

to upper second order. Anomalous interference colors are found in some varieties.

Extinction parallel to the cleavage. The extinction is often wavy on account of the structure.

Orientation.—The cleavage traces are parallel to the fast ray.

Twinning.—Fine polysynthetic twinning in two directions at right angles is found in some sections.

Interference Figure.—Prehnite gives a positive biaxial figure, but the axial angle is variable even in the same specimen. The axial plane is {010}. Dispersion, $r > v$ weak.

Distinguishing Features.—Lawsonite in some occurrences is like prehnite, but the birefringence of lawsonite is considerably lower and the indices of refraction somewhat higher.

Occurrence.—Prehnite is a secondary mineral in cavities and seams of various igneous

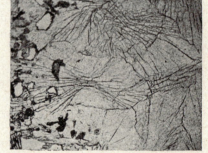

FIG. 318.—(×12) Prehnite showing "bow-tie" structure.

rocks. It is a prominent mineral in amygdaloidal rocks of the Lake Superior copper ores. It is sometimes found in veins. Associated minerals are quartz, calcite, datolite, and zeolites.

Lawsonite

$H_4CaAl_2Si_2O_{10}$ Orthorhombic

$$n_\alpha = 1.665$$
$$n_\beta = 1.674$$
$$n_\gamma = 1.684$$
$$2V = 84°; \text{ Opt. } (+)$$
$$a = \alpha \text{ or } X, \ b = \beta \text{ or } Y, \ c = \gamma \text{ or } Z$$

Color.—Colorless in thin sections. It may be pleochroic in very thick sections.

Form.—Lawsonite occurs in euhedral crystals of varying habit. Sections are usually rhombic (110 $\wedge$ 1$\bar{1}$0 = 67°) or rectangular.

Cleavage good parallel to {010} and {001}, fair parallel to {110}.

Relief rather high, $n >$ balsam.

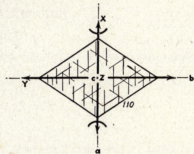

Fig. 319.—Orientation diagram of lawsonite. Section parallel to (001).

Birefringence moderate, $n_\gamma - n_\alpha = 0.019$; so the interference colors range up to second-order blue.

Extinction parallel or symmetrical.

Orientation.—The long diagonal of rhombic sections is parallel to the slower ray.

Fig. 320.—($\times$22) Metacrysts or porphyroblasts of lawsonite in chlorite.

Twinning.—Polysynthetic twinning with {110} as twinplane is rather common. The lamellae are usually thin, may be in either one or two directions, and are sometimes curved.

Interference Figure.—The figure is biaxial positive with a very large axial angle. The axial plane is {010}. Dispersion, $r > v$ strong.

Distinguishing Features.—Clinozoisite somewhat resembles lawsonite, but the anomalous interference colors distinguish it.

Prehnite may also be mistaken for lawsonite, but its birefringence is higher.

Occurrence.—The characteristic occurrence of lawsonite is in metamorphic rocks such as glaucophane schists. It is also found in gabbros and diorites as the result of incipient metamorphism. The type locality is Tiburon Peninsula on San Francisco Bay, California, but it has also been found in Italy, Corsica, and New Caledonia. The usual associates of lawsonite are muscovite, glaucophane, garnet, and sphene.

THE MICA GROUP

The micas constitute a well-defined group of silicates of aluminum together with the alkalies, magnesium, and ferrous

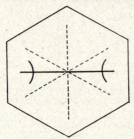 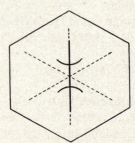

FIG. 321.   FIG. 322.

FIG. 321.—Mica of the first class. The dotted lines represent the "percussion figure."

FIG. 322.—Mica of the second class.

iron. They are characterized by a very perfect cleavage in one direction {001} and by strong birefringence. The extinction angles of sections cut normal to the cleavage are very small or practically zero. The micas are pseudo-hexagonal monoclinic. They may be divided into two classes depending upon whether the optic axial plane is (1) normal to or (2) parallel to {010} (see the above figures). Muscovite and lepidolite belong to the first class, and phlogopite and biotite to the second. All the micas are optically negative.

Lepidomelane is the name given to iron-rich biotite. Fuchsite and mariposite are green chromium-bearing varieties of muscovite. Sericite is a secondary muscovite found in minute shreds and aggregates and formed by hydrothermal alteration.

THE MICA GROUP

Mineral	Chemical composition	n_α	n_β	n_γ	2V
Muscovite.................	KAl	$\begin{cases}1.556 \\ 1.570\end{cases}$	$\begin{matrix}1.587 \\ 1.607\end{matrix}$	$\begin{matrix}1.593 \\ 1.611\end{matrix}$	30–40°
Hydromuscovite	KAl	$\begin{cases}1.535 \\ 1.570\end{cases}$	$\begin{matrix}..... \\ .....\end{matrix}$	$\begin{matrix}1.565 \\ 1.605\end{matrix}$	Small
Lepidolite.................	KLiAl	1.560	1.598	1.605	40°
Phlogopite...............	KMgAl	$\begin{cases}1.551 \\ 1.562\end{cases}$	$\begin{matrix}1.598 \\ 1.606\end{matrix}$	$\begin{matrix}1.598 \\ 1.606\end{matrix}$	0–10°
Biotite....................	KMg,FeAl	$\begin{cases}1.541 \\ 1.579\end{cases}$	$\begin{matrix}1.574 \\ 1.638\end{matrix}$	$\begin{matrix}1.574 \\ 1.638\end{matrix}$	0–25°

MUSCOVITE

$KAl_2(OH)_2(AlSi_3O_{10})$ (inc. Sericite)   Monoclinic
$\angle\beta = 89°54'$

$$n_\alpha = 1.556 \text{ to } 1.570$$
$$n_\beta = 1.587 \text{ to } 1.607$$
$$n_\gamma = 1.593 \text{ to } 1.611$$
$$2V = 30 \text{ to } 40°; \text{ Opt. } (-)$$
$$b = \gamma \text{ or } Z, a \wedge \beta \text{ or } Y, = +1 \text{ to } +3°$$

Color.—Colorless to pale green in thin sections. Some varieties are pleochroic.

Form.—Muscovite usually occurs in thin tabular crystals or in scaly aggregates or shreds. The minutely crystalline variety is called *sericite*.

Cleavage in one direction {001} very perfect.

Relief not marked, $n >$ balsam. On rotation there is some change of relief, fair when the cleavage traces are parallel to the vibration plane of the lower nicol and low in a position at right angles to this.

Birefringence strong, $n_\gamma - n_\alpha = 0.037$ to 0.041; hence the highest interference colors are upper second order. Sections parallel to the cleavage give first-order colors since $n_\gamma - n_\beta = 0.004$ to 0.006.

Extinction.—The extinction is as a rule practically parallel to the cleavage traces, but it is often possible to find angles as high as 2 or 3°.

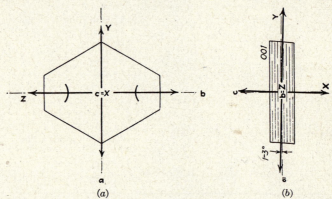

Figs. 323 *a,b.*—Orientation diagrams of muscovite. Sections (*a*) parallel to (001) and (*b*) parallel to (010).

Orientation.—The direction of the cleavage traces is always the slower ray.

Twinning.—Twinning according to the mica law [twin-plane = {110} and composition face = (001)] is fairly common. It may be detected by slight differences in interference colors as well as by extinction angles.

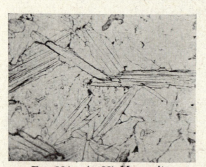

Fig. 324.—(×22) Muscovite.

Interference Figure.—Sections parallel to {001} or cleavage flakes give a biaxial negative figure with a moderate axial angle. Dispersion, $r > v$ weak. The axial plane is normal to {010}.

Distinguishing Features.—Talc is so similar to muscovite and pyrophyllite in its optical properties that it is distinguished with difficulty. The axial angle of talc is smaller. It may be necessary to make a microchemical test for magnesium (use Na_2CO_3 fusion) in order to differentiate them.

Related Minerals.—Hydromuscovite is very similar to the sericite variety of muscovite.

Occurrence.—Muscovite is very common in metamorphic rocks such as phyllites, schists, and gneisses. It is found in some granites and reaches its maximum development in granite pegmatites. It is common as a detrital mineral, especially in arkoses.

Sericite

Sericite occurs in minute shreds and is a secondary mineral formed by hydrothermal alteration of silicates, especially the feldspars. In the opinion of the senior author, it is in all probability a late hydrothermal mineral. Sericite also occurs as a constituent of schists, phyllites, and slates.

Hydromuscovite

For a description of hydromuscovite, see page 360.

Lepidolite

$LiKAl_2(OH,F)_2(Si_2O_5)_2$ Monoclinic

$$n_\alpha = 1.560$$
$$n_\beta = 1.598$$
$$n_\gamma = 1.605$$
$$2V = 40° \pm ; \text{ Opt. } (-)$$

Color.—Colorless in thin sections.

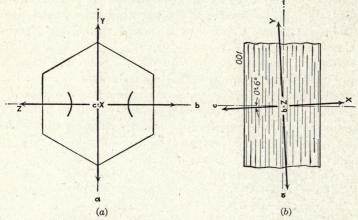

(a) (b)

Figs. 325 *a,b.*—Orientation diagrams of lepidolite. Sections (*a*) parallel to (001) and (*b*) parallel to (010).

Form.—Lepidolite usually occurs in thick tabular or short prismatic pseudo-hexagonal crystals.

Cleavage perfect in one direction {001}.

Relief fair, $n >$ balsam.

Birefringence strong, $n_\gamma - n_\alpha = 0.045$; hence interference colors range up to the middle of the third order. Sections parallel to the cleavage (including cleavage flakes) have weak double refraction ($n_\gamma - n_\beta = 0.007$).

Extinction.—The extinction angle measured against the cleavage traces varies from zero up to a maximum of 6 or 7°.

Orientation.—The direction of the cleavage trace is always the slower ray.

Twinning.—Twinning is common according to the mica law [twin-plane = {110}], the composition face being {001}. Sometimes there are penetration twins.

FIG. 326.—(×25) Lepidolite.

Interference Figure.—The figure is biaxial negative with a moderate axial angle, usually about 40°. Dispersion, $r > v$ weak.

Distinguishing Features.—Lepidolite is very similar to muscovite in its optical properties but has a larger extinction angle. It may be necessary to use some non-optical test to distinguish them. (Lepidolite is easily fusible and gives a lithium flame.)

Related Minerals.—Zinnwaldite is a lithium-iron mica also found in tin-stone veins and granite pegmatites.

Occurrence.—Lepidolite occurs in granite pegmatites, in some high-temperature veins, and occasionally in granites. The usual associates are tourmaline (especially elbaite), albite, topaz, beryl, spodumene, and quartz.

PHLOGOPITE

$KMg_3Al(OH)Si_4O_{10}$

Monoclinic
$\angle\beta = 90°\pm$

$$n_\alpha = 1.551 \text{ to } 1.562$$
$$n_\beta = 1.598 \text{ to } 1.606$$
$$n_\gamma = 1.598 \text{ to } 1.606$$
$$2V = 0 \text{ to } 10°; \text{ Opt. } (-)$$
$$b = \beta \text{ or } Y$$

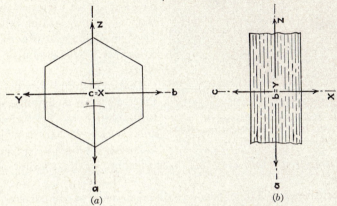

(a)   (b)

Figs. 327 *a,b*.—Orientation diagrams of phlogopite.

Color.—Pale brown to colorless in thin sections. Slightly pleochroic.

Form.—Phlogopite is found in six-sided thick tabular to short prismatic crystals.

Cleavage in one direction {001} as with the other micas.

Relief fair, $n >$ balsam.

Birefringence strong, $n_\gamma - n_\alpha = 0.044$ to 0.047; the maximum interference color is about the middle of the third order. Sections parallel to {001}, which include cleavage flakes, have very weak birefringence since $n_\gamma - n_\beta$ is very small $(0.000x)$.

Fig. 328.—($\times$30) Phlogopite with altered forsterite in metamorphic limestone.

Extinction.—Extinction is usually parallel to the cleavage, but sometimes the extinction angle is as much as 5°.

Orientation.—The cleavage traces are parallel to the slower ray.

Twinning, though often present, is not conspicuous. It may be recognized by differences in interference colors of adjacent parts of a crystal as well as by extinction angles.

Fig. 329.—(×25) Phlogopite with pleochroic halo in schist-gneiss. Note euhedral pyrite.

Interference Figure.— Basal sections, which also include cleavage flakes, give a negative interference figure that is either biaxial with a very small angle or practically uniaxial. Dispersion, $r < v$ weak.

Distinguishing Features.— Phlogopite is distinguished from biotite by lighter color and weaker absorption. Colorless phlogopite is much like muscovite but may be distinguished by its smaller axial angle.

Occurrence.—The characteristic occurrence of phlogopite is in metamorphic limestones. The common associates are chondrodite, spinel, and forsterite. It is also found in a few igneous rocks such as peridotites, derived serpentines, and leucite-bearing rocks.

BIOTITE

$K_2(Mg,Fe)_2(OH)_2(AlSi_3O_{10})$

Monoclinic
$\angle \beta = 90°$

$$n_\alpha = 1.541 \text{ to } 1.579$$
$$n_\beta = 1.574 \text{ to } 1.638$$
$$n_\gamma = 1.574 \text{ to } 1.638$$
$$2V = 0 \text{ to } 25°; \text{ Opt. } (-)$$
$$b = \beta \text{ or } Y, c \wedge \alpha \text{ or } X = 3° \pm$$

Color.—Brown, yellowish brown, reddish brown, olive green, or green in thin sections. Pleochroic. The absorption is stronger when the cleavage traces are parallel to the vibration plane of the lower nicol.

Form.—Common in euhedral six-sided crystals that are usually tabular in habit; also in lamellar aggregates. The plates are sometimes bent.

Inclusions.—Inclusions of such minerals as zircon surrounded by pleochroic haloes are fairly common in biotite.

Cleavage perfect in one direction {001}. Sections cut parallel to {001} do not show any cleavage. In schistose rocks these sections predominate.

Relief fair, n > balsam.

Birefringence strong, $n_\gamma - n_\alpha = 0.033$ to 0.059; interference colors range up to second-order red, but the color of the mineral

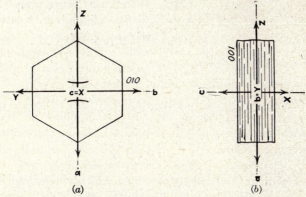

FIGS. 330 *a,b.*—Orientation diagrams of biotite. Sections (*a*) parallel to (001) and (*b*) parallel to (010).

may mask the interference color. The birefringence of sections parallel to {001} is practically nil since $n_\gamma - n_\beta = 0.000x$.

Extinction usually parallel to the cleavage traces, but the extinction angle may be as much as 3° in some sections. Bent plates have wavy extinction. Near the extinction position a peculiar crinkly appearance is usually noticed in biotite.

Orientation.—The direction of the cleavage traces is always the slower ray.

Twinning.—Twinning according to the mica law [twin-plane = {110}] may be present.

Interference Figure.—Sections parallel to {001}, including cleavage flakes, give a negative biaxial figure with a very small axial angle. The axial plane is usually parallel to {010}, but in one variety (anomite) the axial plane is normal to {010}. Dispersion, $r > v$ or $r < v$ weak.

Distinguishing Features.—Biotite is distinguished from phlogopite by its darker color and stronger absorption. From ordinary brown hornblende it is distinguished by the smaller extinction

angle and difference in cleavage. Lamprobolite also has the typical amphibole cleavage. Tourmaline has strong absorption when elongation of the crystals is normal to the vibration plane of the lower nicol.

Related Minerals.—Lepidomelane resembles biotite but has higher indices and larger axial angle. Manganophyll is a manganian biotite found in

Fig. 331.—(×15) Biotite in gneiss showing pleochroic halos.

metamorphic dolomite at Långban, Sweden.

Alteration.—Biotite is often more or less altered to chlorite. It may also alter to vermiculite. In the opinion of E. W. Galliher, detrital biotite is the source of practically all glauconite.

Occurrence.—Biotite is a widely distributed and common mineral. It occurs in igneous rocks of nearly all types. It is also a prominent constituent of schists and gneisses and may be found in contact-metamorphic zones. Biotite is common in detrital sediments. It is often bleached or otherwise altered.

CHLORITE GROUP

The chlorites are basic magnesium-iron aluminum silicate minerals of micaceous appearance. They are usually green, often pleochroic, monoclinic pseudo-hexagonal with moderate refringence and weak to rather weak birefringence. Cleavage is perfect in one direction parallel to (001).

Chlorites may be divided into two general subgroups: (1) orthochlorites, which are low in iron and are generally well crystallized, and (2) leptochlorites high in iron and not so well crystallized.

Orthochlorites include clinochlore, pennine, and prochlorite; the most important leptochlorite is chamosite.

Descriptions of the four minerals mentioned are given here, but it must be recognized that there are many other chlorites. It is often difficult to determine the individual mineral, and so it is often necessary simply to classify it as "chlorite."

Clinochlore

$Mg_5(Al,Fe)(OH)_8(Al,Si)_4O_{10}$

Monoclinic
$\angle\beta = 89°40'$

$$n_\alpha = 1.571 \text{ to } 1.588$$
$$n_\beta = 1.571 \text{ to } 1.588$$
$$n_\gamma = 1.576 \text{ to } 1.597$$
$$2V = 0 \text{ to } 50°; \text{ Opt. } (+)$$
$$b = \beta \text{ or } Y, c \wedge \gamma \text{ or } Z = 2 \text{ to } 9°$$

Color.—Colorless to green in thin sections. Pleochroic with absorption: α or X and β or Y $> \gamma$ or Z.

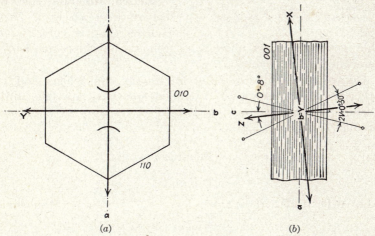

(a) $\qquad\qquad\qquad\qquad$ (b)

FIGS. 332 a,b.—Orientation diagram of clinochlore. Sections (a) parallel to (001) and (b) parallel to (010).

Form.—The crystal habit varies from thin to thick tabular with pseudo-hexagonal outlines. Crystals are often bent.

Cleavage perfect in one direction parallel to {001}.

Relief fair, $n >$ balsam.

Birefringence weak to rather weak, $n_\gamma - n_\alpha = 0.004$ to 0.011.

Extinction.—The maximum extinction angle measured from cleavage traces varies from 2 to 9°. Basal sections are practically isotropic.

Orientation.—Crystals showing cleavage are usually length-fast.

Twinning.—Polysynthetic twinning is common, according to the mica law.

Interference Figure.—The interference figure is biaxial positive

FIG. 333.—(×15) Chlorite (clino-chlore) showing a euhedral crystal in groundmass of chlorite of a different kind.

with a variable axial angle. The axial plane is {010}. Dispersion, $r < v$.

Distinguishing Features.—Clinochlore is distinguished from other chlorites by the oblique extinction and from pennine by greater bire-fringence and distinctly biaxial character.

Related Minerals.—Leuchtenbergite is an iron-free chlorite that is colorless in thin sections. Kotschubeite is a chromian clinochlore frequently associated with chromite.

Occurrence.—Clinochlore occurs in chlorite schists and in other metamorphic rocks. Common associates are talc, antigorite, chondrodite, and phlogopite. It is also an alteration product of other silicates.

Pennine

(Penninite)

$Mg_5(Al,Fe)(OH)_8(Al,Si)_4O_{10}$ Monoclinic

$$n_\alpha = 1.575 \text{ to } 1.582$$
$$n_\beta = 1.576 \text{ to } 1.582$$
$$n_\gamma = 1.576 \text{ to } 1.583$$
$$2V = 0 \text{ to } 20°; \text{ Opt. } (+) \text{ or } (-)$$

Color.—Green or greenish in thin sections. Pleochroic from green to nearly colorless: occasionally from green to brownish red.

Form.—Pennine usually occurs in six-sided crystals of thick tabular habit.

Cleavage perfect in one direction parallel to {001}.

Relief fair, $n >$ balsam.

Birefringence very weak, $n_\gamma - n_\alpha = 0.001$ to 0.004; the interference colors are often anomalous "Berlin blue," a color not found on the interference color chart.

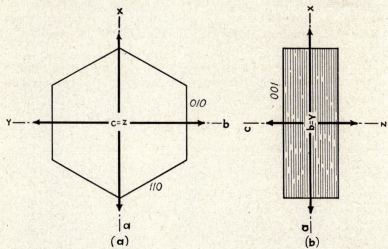

FIGS. 334 *a,b.*—Orientation diagrams of pennine. Sections (*a*) normal to the *c*-axis and (*b*) parallel to (010).

Extinction parallel or almost parallel to cleavage traces and to crystal outlines.

Orientation.—The cleavage traces are parallel to the faster ray.

Twinning.—Twinning parallel to {001} with (001) as the composition face is so characteristic that this mode of twinning is known as the *pennine law.* Since the extinction is practically parallel this twinning is not easily recognized in thin sections.

Interference Figure.—The interference figure is biaxial with such a small axial angle

FIG. 335.—($\times$50) Chlorite (pennine) showing pleochroism and absorption.

that it appears to be uniaxial. The optical sign is usually positive, but sometimes it is negative. The axial plane is usually parallel to {010}.

Distinguishing Features.—Pennine is distinguished from most of the other chlorites by parallel extinction, very small axial angle, and anomalous interference colors. The indices of refraction are lower than those of prochlorite.

Related Minerals.—Kämmererite is a chromian pennine very similar to kotschubeite. Pseudophite is a compact tough variety of pennine. It is one of the pseudojades known as "Styrian jade."

Occurrence.—Pennine usually occurs as an alteration product of other silicates such as garnet. The most typical specimens are found in seams and cavities.

Prochlorite

(Ripidolite)   Monoclinic

$$n_\alpha = 1.588 \text{ to } 1.658$$
$$n_\beta = 1.589 \text{ to } 1.667$$
$$n_\gamma = 1.599 \text{ to } 1.667$$
$$2V = 0 \text{ to } 30°; \text{ Opt. } (+)$$

Color.—Green or greenish in thin sections. Pleochroism weak.

Form.—Prochlorite usually occurs in scaly masses. It is also frequently found in vermicular crystals with hexagonal cross sections and in fan-shaped crystal aggregates.

Cleavage in one direction parallel to {001} as in the other chlorites.

Relief fair to moderately high, $n >$ balsam.

Birefringence usually weak but varies from very weak to rather weak, $n_\gamma - n_\alpha = 0.001$ to 0.011.

Extinction parallel to almost parallel.

Orientation.—Cleavage traces are parallel to the faster ray.

Interference Figure.—The interference figure is usually difficult to obtain. When found, it is usually biaxial positive with a very small axial angle, often practically uniaxial. The axial plane is parallel to {010}. Dispersion, $r < v$.

Distinguishing Features.—Prochlorite is distinguished from clinochlore and pennine by higher indices of refraction.

Occurrence.—Prochlorite is the principal constituent of some chlorite schists often accompanied by magnetite. It also is found as an alteration product of other silicates. A characteristic occurrence of prochlorite is in quartz veins with adularia, albite, sphene, etc.

Chamosite

$Fe_3^{II}Al_2Si_2O_{10}.3H_2O$ Monoclinic(?)

$$n\ ca. = 1.635$$
$$2V\ small;\ Opt.\ (-)$$

Color green, greenish gray, gray, pale brown to almost colorless in thin sections. Some sections show slight pleochroism.

Form usually oolitic with pseudo-spherulitic structure (concentric instead of fibrous elements). Subhedral crystals of thick tabular habit are occasionally found. The ooliths often have a portion of a chamosite crystal as a nucleus. Chamosite is sometimes massive with aggregate structure.

Fig. 336.—(×28) Oolitic chamosite with siderite and detrital minerals.

Cleavage in one direction, but not as perfect as in the micas. The ooliths often show concentric parting.

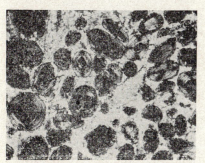

Fig. 337.—(×30) Chamosite ooliths showing pseudo-spherulitic structure in a matrix of calcite. (×nicols.)

Relief moderate, $n >$ balsam. The index of refraction is in the neighborhood of 1.635.

Birefringence nil to weak, up to about 0.007 to 0.008. Interference colors are not appreciably anomalous.

Orientation.—Both cleavage traces and the concentric layers of the ooliths are length-slow.

Distinguishing Features.—Chamosite is distinguished from the orthochlorites by higher index of refraction and from glauconite by lower birefringence. Chamosite often resembles oolitic collophane, but the presence of chamosite crystals and the slightly higher index of refraction will usually distinguish it.

Related Minerals.—The leptochlorites, delessite and thuringite, are similar to chamosite, but the former occurs in true spherulites in amygdaloidal rocks and the latter in schists. Greenalite, an iron silicate occurring in amorphous granules in cherts of the Lake Superior region, also resembles chamosite.

Occurrence.—Chamosite is a prominent constituent of oolitic sedimentary iron ores that are prominent in the Jurassic of England, where they are usually known as *ironstones* and are commercially important low-grade ores. According to Hallimond, they are for the most part of marine origin, but some are probably fresh water. The usual associates of chamosite are siderite, calcite, collophane, pyrite, and various detrital minerals.

These chamositic ores are also found in Scotland, Lorraine, southern Sweden, Bohemia, and Newfoundland.

Chloritoid

$H_2(Fe,Mg,Mn)Al_2SiO_7$ (Ottrelite)   Monoclinic

$$n_\alpha = 1.715 \text{ to } 1.724$$
$$n_\beta = 1.719 \text{ to } 1.726$$
$$n_\gamma = 1.731 \text{ to } 1.737$$
$$2V = 36 \text{ to } 63°; \text{ Opt. } (+)$$
$$b = \beta \text{ or } Y, c \wedge \gamma \text{ or } Z = +3 \text{ to } +21°$$

Color.—Green, greenish gray to colorless. Usually more or less pleochroic.

Form.—Chloritoid, one of the group known as *brittle micas*, commonly occurs in pseudo-hexagonal tabular crystals. Inclusions are often present. It often shows a kind of "hourglass" structure.

Cleavage perfect in one direction {001} and imperfect parallel to {110}.

Relief high, $n >$ balsam.

Birefringence weak to moderate, $n_\gamma - n_\alpha = 0.013$ to 0.016. Basal plates are practically isotropic.

Extinction.—Almost parallel up to *ca.* 20°.

Orientation.—The crystals are length-fast.

Twinning.—Polysynthetic twins after the mica law are very common.

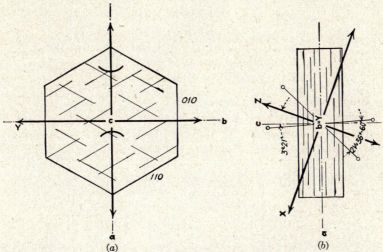

FIGS. 338 *a,b.*—Orientation diagrams of chloritoid. Sections (*a*) normal to the *c*-axis and (*b*) parallel to (010).

Interference Figure.—The figure is biaxial positive with a moderate to rather large axial angle.   The axial plane is {010}. Dispersion, $r > v$ or $r < v$.

Distinguishing Features.—Chloritoid somewhat resembles some of the chlorites, but the relief is much higher and the cleavage less perfect.

Related Minerals.—Ottrelite usually treated as a variety of chloritoid, is considered by Mèlon to be a distinctive mineral.

FIG. 339.—(×12) Chloritoid in schist.

Occurrence.—Chloritoid occurs in metamorphic rocks such as mica schists and phyllites as metacrysts.

TALC

$Mg_3(OH)_2(Si_2O_5)_2$ Monoclinic
$\angle\beta = (?)$

$$n_\alpha = 1.538 \text{ to } 1.545$$
$$n_\beta = 1.575 \text{ to } 1.590$$
$$n_\gamma = 1.575 \text{ to } 1.590$$
$$2V = 6 \text{ to } 30°; \text{ Opt. } (-)$$
$$a = \beta \text{ or } Y,\ b = \gamma \text{ or } Z,\ c = \alpha \text{ or } X$$

Color.—Colorless in thin sections.

Form.—Talc occurs in coarse to fine platy or fibrous aggregates that often have a more or less parallel arrangement. Shreds and plates are often bent. Euhedral crystals of talc are unknown.

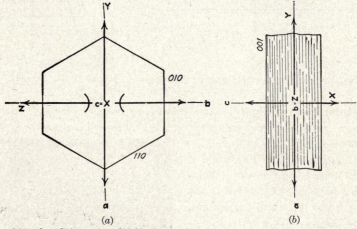

Figs. 340 *a,b.*—Orientation diagrams of talc. Sections parallel to (*a*) (001) and (*b*) (010).

Cleavage perfect in one direction {001}.

Relief fair, $n >$ balsam.

Birefringence very strong, $n_\gamma - n_\alpha = 0.030$ to 0.050; the maximum interference colors are upper third order. Sections parallel to the cleavage give very low first-order gray colors since $n_\gamma - n_\beta$ is almost nil (0.000*x*).

Extinction.—The extinction is parallel to the cleavage traces in most sections; in a few sections the extinction is 2 or 3°; hence talc is probably monoclinic.

Orientation.—Cleavage traces and shreds are length-slow as in muscovite.

Interference Figure.—Cleavage flakes give a biaxial negative figure with a small axial angle. Dispersion, $r > v$ distinct.

Distinguishing Features.—Talc greatly resembles muscovite and pyrophyllite, but may often be distinguished by the smaller axial angle provided an interference figure can be obtained.

It may be necessary to make a chemical or microchemical test in order to prove the identity of talc. The association with other magnesium minerals indicates the presence of talc rather than muscovite or sericite.

Occurrence.—Talc is the principal constituent of talc schists and soapstones. It is often a hydrothermal mineral formed at the expense of antigorite and tremolite in shear zones of serpentines. Dolomite and magnesite are frequent associates.

Pyrophyllite

$Al_2(OH)_2Si_4O_{10}$ Monoclinic(?)

$$n_\alpha = 1.552$$
$$n_\beta = 1.588$$
$$n_\gamma = 1.600$$
$$2V = 53 \text{ to } 60°; \text{ Opt. } (-)$$
$$a = \gamma \text{ or } Z, \ b = \beta \text{ or } Y, \ c = \alpha \text{ or } X$$

Color.—Colorless in thin sections.

Form.—Pyrophyllite occurs in subhedral crystals that are tabular parallel to {010} and much elongated. The crystals are usually curved and distorted. A radial structure is common. It also occurs in fine aggregates.

Cleavage perfect in one direction {001}.

Relief rather low to moderate, $n >$ balsam.

Birefringence strong,

$$n_\gamma - n_\alpha = 0.048;$$

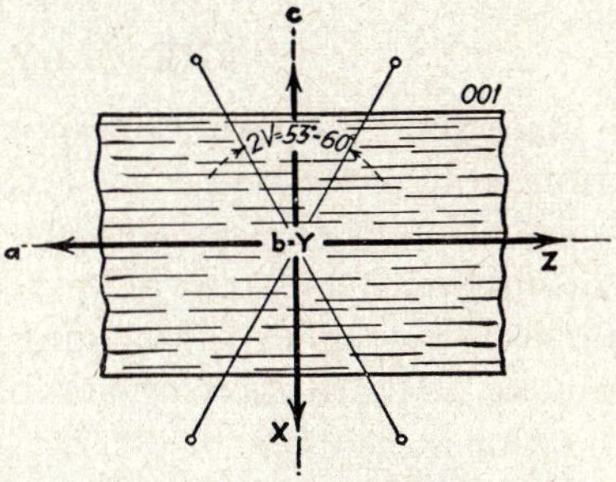

Fig. 341.—Orientation diagram of pyrophyllite. Section parallel to (010).

the maximum interference color is upper third order. Sections parallel to the cleavage give gray or first-order colors since $n_\gamma - n_\beta = 0.012$.

Extinction parallel or almost parallel to cleavage traces and parallel to the elongate sections.

Orientation.—Cleavage traces are parallel to the slower ray. Elongate sections not showing cleavage are length-slow.

Twinning.—Twinning like that found in the micas is present but is not well defined.

Interference Figure.—Sections parallel to {001} or cleavage flakes give a biaxial negative figure with a rather large axial angle. The axial angle is parallel to {010} or the length of the crystals.

Fig. 342.—(×50) Pyrophyllite.

Distinguishing Features.—Pyrophyllite usually has a peculiar elongate tabular habit. Muscovite and talc greatly resemble pyrophyllite. Talc has a much smaller axial angle. Microcrystalline pyrophyllite is very difficult to distinguish from sericite or talc by optical means; it may be necessary to make chemical or microchemical tests.

Occurrence.—Pyrophyllite occurs in metamorphic rocks and has often developed as a hydrothermal alteration product. Common associates are andalusite, sillimanite, kyanite, lazulite, and alunite.

THE CLAY MINERALS

The minerals that characterize clays are most widely distributed in sedimentary rocks, but in addition the same minerals may be found occasionally in veins or as alteration products of igneous and metamorphic rocks. The most common clay minerals belong to the kaolin, montmorillonite, and hydromica groups. All are finely crystalline or meta-colloidal and are likely to occur in flakelike or dense aggregates of varying types.

The kaolin group includes the minerals kaolinite, anauxite, dickite, nacrite, halloysite, hydrohalloysite, and allophane. Of these, kaolinite, dickite, and nacrite have the same composition $(Al_2O_3.3SiO_2.2H_2O)$ but differ optically and also in internal structure, as shown by X-ray diffraction. Anauxite (Al_2O_3.-$3SiO_2.2H_2O$) corresponds to kaolinite in internal structure and is only slightly different optically. Halloysite ($Al_2O_3.2SiO_2$.-$2H_2O$) and hydrated halloysite differ somewhat optically and also structurally. Allophane is a solidified gel ($Al_2O_3.xSiO_2$.-nH_2O) with an index of refraction lower than the indices of the other minerals of the group.

Kaolinite is the most common of the minerals of the kaolin group. It occurs in large quantities, both as a residual and as a transported mineral, as, for example, in the white clays of the southeastern United States. It is usually formed by the decomposition *in situ* of rocks containing feldspars. Dickite, although less common, is frequently better crystallized and may occur in pseudo-hexagonal plates. It is commonly formed by hydrothermal solutions in veins or dikes and occasionally occurs associated with sulfids in ore deposits. Halloysite is found as a porcelainlike mass or fine white powder in shales and sandstones and as a replacement of limestone. It may occur independently but is most frequently associated with kaolinite, alunite, and various forms of hydrous aluminum oxides.

Montmorillonite is one of the most widely distributed of the clay minerals. It occurs as an alteration product of volcanic ash and tuff. The rock bentonite, common in the western United States, is composed largely of the mineral montmorillonite. Montmorillonite is also found in minor amounts as an alteration product in pegmatite dikes. The montmorillonite group consists of montmorillonite [$(Mg,Ca)O.Al_2O_3.5SiO_2.nH_2O$] beidellite ($Al_2O_3.3SiO_2.nH_2O$), nontronite [$(Al,Fe)_2O_3.3SiO_2.nH_2O$] and saponite ($2MgO.3SiO_2.nH_2O$).

Hydromica represents a poorly defined mineral group long recognized as an intermediate product formed during the alteration of feldspathic minerals to kaolinite. It is likely to occur in shale or other argillaceous beds where it may be associated with kaolinite and mixed with fine detrital fragments of other minerals. The general formula $(OH)_4K_\gamma(Al_4Fe_4Mg_4Mg_6)(Si_{8-\gamma}.-Al_\gamma)O_{20}(\gamma = 1$ to 1.5) suggested by Grim[1] indicates the complex chemical character of hydromica.

---

[1] The name *illite* has been applied to this group, but it is believed that *hydromuscovite*, as described by A. Johnstone (*Quar. Jour. Geol. Soc.*, vol. 45, p. 363, 1889) merits priority.

CLAY MINERALS

	Chemical composition	Crystal system	$n\alpha$	$n\beta$	$n\gamma$	$n\gamma - n\alpha$
Kaolin group						
Kaolinite	$Al_2O_3.2SiO_2.2H_2O$	Monoclinic	1.561	1.565	1.566	0.005
Anauxite	$Al_2O_3.3 \pm SiO_2.2H_2O$	Monoclinic		(see kaolinite)		
Dickite	$Al_2O_3.2SiO_2.2H_2O$	Monoclinic	1.560	1.562	1.566	0.006
Nacrite	$Al_2O_3.2SiO_2.2H_2O$	Monoclinic	1.557	1.562	1.563	0.006
Halloysite	$Al_2O_3.2 \pm SiO_2.2 + H_2O$	Aggregates		$n = 1.549\text{-}1.561$		0.001
Hydrohalloysite	$Al_2O_3.2 \pm SiO_2.4 + H_2O$	Aggregates		$n = 1.526\text{-}1.542$		
Allophane	$Al_2O_3.xSiO_2.nH_2O$	Amorphous		$n = 1.47\text{-}1.49$		
Montmorillonite group						
Montmorillonite	$(Mg,Ca)O.Al_2O_3.5SiO_2.nH_2O$	Monoclinic	1.492		1.513	0.021
Beidellite	$Al_2O_3.3SiO_2.nH_2O$	Monoclinic	1.517		1.549	0.032
Nontronite	$Fe_2O_3.3SiO_2.nH_2O$	Monoclinic	1.580		1.615	0.035
Saponite	$2MgO.3SiO_2.nH_2O$	Monoclinic	1.479-1.490	1.510-1.525	1.511-1.527	0.032-0.037
Hydromica group						
Hydromuscovite	$KAl_2(OH)_2[AlSi_3(O,OH)_{10}]$	Monoclinic	1.535-1.57		1.565-1.605	0.030-0.035

The significant optical properties of the clay minerals may be conveniently outlined for the three groups from the standpoint of microscopic classification: (1) The kaolin group is characterized by weak birefringence and indices of refraction approximating those of quartz. (2) The montmorillonite group has a comparatively high birefringence but indices of refraction for the most part lower than Canada balsam. (3) The hydromica group is characterized by a comparatively high birefringence with indices of refraction above Canada balsam.

Potash-bearing clay, probably hydromica, is frequently found as an alteration product of old volcanic ash, particularly in the Ordovician of the eastern United States, or at times as a gouge clay in ore deposits.

A strong artificial illumination is advisable in determining the birefringence of the clay minerals. A summary of the optical properties of the more important clay minerals based upon determinations either by C. S. Ross or R. E. Grim is given in the foregoing tabulation.

The most important clay minerals exhibit distinguishing features in thin sections. A number, however, present problems in identification best solved by coordinated optical and X-ray methods.

KAOLINITE

$Al_2O_3.2SiO_2.2H_2O$ Monoclinic

$$n_\alpha = 1.561$$
$$n_\beta = 1.565$$
$$n_\gamma = 1.566$$
$$2V \text{ variable; Opt. } (-)$$
$$b = \gamma \text{ or } Z, c \wedge \alpha \text{ or } X = 1 \text{ to } 3\tfrac{1}{2}°$$

Color.—Colorless to pale yellow.

Form.—Kaolinite occurs in fine mosaiclike masses of crystals, in veinlets replacing feldspars and other minerals, and in scalelike individuals. Occasionally small plates show accordionlike outlines.

Cleavage perfect in one direction parallel to {001}.

Relief low, $n >$ balsam.

Birefringence weak, $n_\gamma - n_\alpha = 0.005$. In normal sections kaolinite gives gray and white interference colors.

Extinction.—The angle of extinction on (010) against the base is 1 to $3\frac{1}{2}°$.

Orientation.—The cleavage traces and crystals are length-slow.

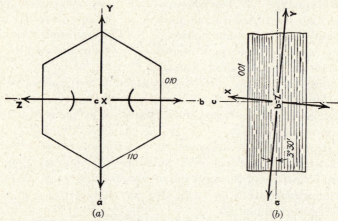

FIGS. 343 *a,b.*—Orientation diagrams of kaolinite. Sections (*a*) normal to the *c*-axis and (*b*) parallel to (010).

Twinning.—Minute crystals of kaolinite do not appear to show twinning.

Interference Figure.—Kaolinite is ordinarily too fine-grained to give an interference figure. The axial plane is normal to {010} and nearly parallel to {100}.

FIG. 344.—(×85) Kaolinite from near St. Etienne, France. (*Courtesy of U. S. Geological Survey.*)

FIG. 345.—(×85) Kaolinite (anauxite) from Bilin, Bohemia. (*Courtesy of U. S. Geological Survey.*)

Distinguishing Features.—It is distinguished by low relief and weak birefringence. From dickite it is distinguished largely by its smaller extinction angle.

Occurrence.—Kaolinite is found as a weathering product of igneous and metamorphic rocks. It is produced particularly by the decomposition of feldspars. It occurs as a prominent clay mineral in sedimentary beds.

Dickite

$Al_2O_3.2SiO_2.2H_2O$ Monoclinic

$$n_\alpha = 1.560$$
$$n_\beta = 1.562$$
$$n_\gamma = 1.566$$
$$2V = 52 \text{ to } 80°; \text{ Opt. } (+)$$
$$b = Z, \ c \wedge X = 15 \text{ to } 20°$$

Color.—Colorless to pale yellow.

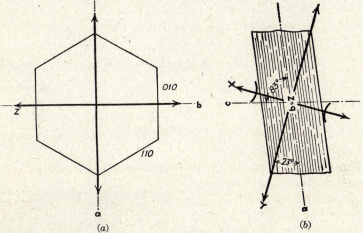

(a)   (b)

FIGS. 346 *a,b.*—Orientation diagrams of dickite. Sections (*a*) normal to the c-axis and (*b*) parallel to (010).

Form.—Dickite occurs in small pseudo-hexagonal flakelike crystals.

Cleavage.—Perfect in one direction parallel to {001}.

Relief low, $n >$ balsam.

Birefringence weak, $n_\gamma - n_\alpha = 0.006$; the maximum interference colors are middle first order.

Extinction.—Angle of extinction on {010} against base varies from 15 to 20°.

Orientation.—The cleavage traces and crystals are length-slow.

Twinning.—No apparent twinning.

Interference Figure.—Ordinary thin sections yield poor

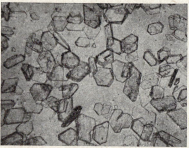

interference figures or none at all. The axial plane is normal to {010}. Dispersion, $r < v$.

Distinguishing Features.—Dickite is distinguished from kaolinite by slightly higher birefringence, larger crystals, and a larger extinction angle.

Occurrence.—The mineral is commonly found associated with metallic minerals in ore deposits. It is usually pro-

FIG. 347.—($\times$260) Crystals of dickite, Red Mountain, Colorado. (*Courtesy of U. S. Geological Survey.*)

duced by hydrothermal action and may occur as a replacement of quartz in quartzite.

Halloysite

$Al_2O_3.2SiO_2.2H_2O$

Aggregates
(Crystal System
Unknown)

$$n = 1.549 \text{ to } 1.561$$

Color.—Colorless.

Form.—Halloysite occurs in extremely fine-grained or colloform masses and commonly shows shatter cracks.

Relief low, n slightly $>$ balsam.

Birefringence very weak, almost isotropic.

Distinguishing Features.—Halloysite is distinguished by extremely weak birefringence and an index of refraction almost equal to balsam and shatter cracks. It is commonly associated with other clay minerals.

FIG. 348.—($\times$27) Shatter cracks in halloysite. ($\times$ nicols.)

Occurrence.—The mineral occurs in altered areas in limestone associated with diaspore, alunite, or gibbsite; in clay beds associated with kaolinite; and probably in extremely weathered portions of some shales.

MONTMORILLONITE

$(Mg,Ca)O.Al_2O_3.5SiO_2.nH_2O$ Monoclinic

$$n_\alpha = 1.492$$
$$n_\beta = n_\gamma = 1.513$$
$$2V = 10 \text{ to } 25°; \text{Opt.} (-)$$

Color.—Pale pink, greenish, or colorless.

Form.—Massive, claylike microcrystalline aggregates in the shape of shards. Nearly always in extremely fine scale-like crystals. Crystals of this type have been described by E. T. Wherry as a one-dimensional colloid since they are usually so thin.

Relief rather low, $n <$ balsam.

Birefringence moderate, $n_\gamma - n_\alpha = 0.021$. Although the birefringence is moderate, the crystals are usually so thin

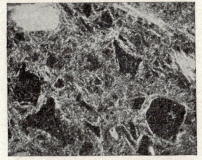

Fig. 349.—(×20) Montmorillonite, more or less altered glass shards, and feldspar fragments in bentonite. (× nicols.)

that interference colors in thin sections seldom go above the second order.

Interference Figure.—Individual crystals are usually so small and thin that figures cannot be obtained.

Distinguishing Features.—The most characteristic feature of montmorillonite is microcrystalline aggregates in the shape of shards.

Occurrence.—Montmorillonite is the chief constituent of bentonite, which is altered volcanic ash. It also occurs in fuller's earth as a primary constituent and has been found as an alteration material in pegmatite dikes.

Hydromuscovite

(Hydromica)

(Illite)

$KAl_2(OH)_2[AlSi_3(O,OH)_{10}]$ Monoclinic(?)

$$n_\alpha = 1.535 \text{ to } 1.570$$
$$n_\gamma = 1.565 \text{ to } 1.605$$
$$2V \text{ small; Opt. } (-)$$
$$\alpha \text{ or } X = \perp \text{ plane of } a \text{ and } c.$$

Color.—Colorless to yellowish brown.

Form.—Hydromuscovite occurs in irregular matted flakes that may be intercalated with flakes of kaolinite.

Relief low, $n >$ balsam.

Birefringence rather strong, $n_\gamma - n_\alpha = 0.030$ to 0.035, but the small, thin crystals may not yield colors above the second order.

Distinguishing Features.—Hydromuscovite occurs in matted small flakes resembling kaolinite and montmorillonite but

Fig. 350.—Kaolinite prisms with intergrown hydromuscovite plates. (*After S. G. Galpin.*)

is distinguished from montmorillonite by higher indices of refraction and from kaolinite by a higher birefringence. It is distinguished from muscovite by a lower axial angle.

Occurrence.—Hydromuscovite occurs as an intermediate product during the alteration of feldspathic minerals, biotite, muscovite, and other constituents of shale or soil.

THE SERPENTINE MINERALS

Serpentine is here considered to be a rock made up of various magnesium minerals. Of these, antigorite generally predominates but there may also be serpophite (recently named and described by Lodochnikov) and chrysotile.

ANTIGORITE

(Serpentine in part)

$H_4Mg_3Si_2O_9$ Orthorhombic

$$n_\alpha = 1.555 \text{ to } 1.564$$
$$n_\beta = 1.562 \text{ to } 1.573$$
$$n_\gamma = 1.562 \text{ to } 1.573$$
$$2V = 20 \text{ to } 90°; \text{ Opt. } (-)$$
$$a = \beta \text{ or } Y, \; b = \alpha \text{ or } X, \; c = \gamma \text{ or } Z$$

Color.—Colorless to pale green in thin sections.

Form.—Antigorite occurs in anhedral crystals or aggregates of fibrolamellar structure. It often occurs as pseudomorphs after pyroxene (bastite), olivine, etc.

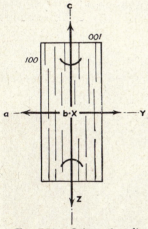

Fig. 351.—Orientation diagram of antigorite. Section parallel to (010).

Fig. 352.—(×20) Antigorite in serpentine. (× nicols.)

Relief rather low, $n >$ balsam.

Birefringence weak, $n_\gamma - n_\alpha = 0.007$ to 0.009; the maximum interference color is first-order yellow. This yellow is slightly anomalous since it has a greenish tinge.

Extinction parallel.

Orientation.—The crystals are length-slow.

Interference Figure.—The figure is biaxial negative with variable axial angle. The axial plane is {100}. Dispersion, $r > v$ weak.

Distinguishing Features.—Chrysotile is distinguished from its dimorph antigorite by the fine fibrous structure. Antigorite usually shows aggregate structure and is in practically all cases an alteration product of some other silicate mineral. Serpophite has lower birefringence than antigorite and shows little or no form or structure.

Occurrence.—Antigorite is the main constituent of serpentine, a metamorphic rock. It has been formed from olivine, enstatite, augite, etc., by hydrothermal alteration. Common associates are chrysotile, talc, magnetite, chromite, and picotite.

Serpophite

(Serpentine in part)

$H_4Mg_3Si_2O_9$ Amorphous
(Mineraloid)

$$n = 1.50–1.57$$

Color.—Colorless to very pale green in thin sections.

Form.—Massive, almost structureless, often occurring in cores surrounded by antigorite.

FIG. 353.—(×25) Serpophite (clear smooth areas) and antigorite (rough gray) with secondary magnetite in serpentine.

FIG. 354.—Serpophite (black or dark gray) with lenticular antigorite in serpentine. (The same spot as Fig. 353.) (× nicols.)

Cleavage absent.

Relief very low, $n >$ balsam, $n <$ antigorite.

Birefringence nil to very weak, not over 0.003. Interference colors: either none or very low first-order gray.

Extinction.—The extinction of serpophite is more or less wavy and usually rather indefinite.

Distinguishing Features.—Serpophite is distinguished from the other serpentine minerals by its very weak birefringence and its lack of structure.

Occurrence.—Serpophite is one of the constituents of serpentine (used here as a rock name). It is usually intimately associated with antigorite.

Chrysotile

$H_4Mg_3Si_2O_9$ Orthorhombic

$$n_\alpha = 1.493 \text{ to } 1.546$$
$$n_\beta = 1.504 \text{ to } 1.550$$
$$n_\gamma = 1.517 \text{ to } 1.557$$
$$2V = 0 \text{ to } 50°; \text{ Opt. } (+)$$

Color.—Colorless in thin sections.

Form.—Chrysotile occurs in cross-fiber veinlets.

Relief low, n slightly greater than balsam.

Birefringence moderate, $n_\gamma - n_\alpha = 0.011$ to 0.014; the maximum interference color is bright yellow of the first order.

Extinction parallel.

Orientation.—The fibers are length-slow.

Distinguishing Features.— The other forms of asbestos (tremolite, anthophyllite, and crocidolite) all have higher

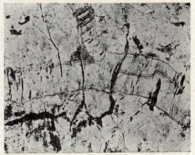

Fig. 355.—(×30) Chrysotile veinlets in serpentine with secondary magnetite.

indices of refraction than chrysotile. Tremolite has oblique extinction.

Occurrence.—Chrysotile usually occurs in veinlets in serpentine that consists largely of the mineral antigorite.

Iddingsite

MgO.Fe$_2$O$_3$.3SiO$_2$.4H$_2$O   Orthorhombic

$$n_\alpha = 1.674 \text{ to } 1.730$$
$$n_\beta = 1.715 \text{ to } 1.763$$
$$n_\gamma = 1.718 \text{ to } 1.768$$
$$2V = 25 \text{ to } 60°; \text{ Opt. } (+) \text{ or } (-)$$
$$a = \alpha \text{ or } X, b = \beta \text{ or } Y, c = \gamma \text{ or } Z$$

Color.—Brown in thin sections. Pleochroism slight to distinct. Absorption: γ or Z > β or Y > α or X.

Form.—Iddingsite, as far as known, always occurs as partial or complete pseudomorphs after olivine. It shows a lamellar structure.

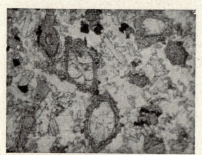

FIG. 356.—(×16) Iddingsite alteration border around olivine phenocrysts.

Cleavage in three directions {100}, {001}, {010} at right angles. Of these {100} has the most perfect cleavage.

Relief high, n > balsam.

Birefringence strong, $n_\gamma - n_\alpha = 0.038$ to 0.044; the maximum interference color should be in the third order, but the color of the mineral modifies or masks the interference color.

Extinction parallel to the cleavage traces.

Interference Figure.—The figure is biaxial, either positive or negative, with a moderate axial angle. The axial plane is {010}. Dispersion, $r > v$ or $r < v$ strong.

Distinguishing Features.—The reddish brown color and lamellar structure together with the mode of occurrence are distinctive for iddingsite.

Alteration.—Iddingsite is sometimes found more or less altered to limonite or indefinite hydrous iron oxids.

Occurrence.—Iddingsite is found in basalts and basalt porphyries as an alteration product of olivine. According to Ross and Shannon, it is a deuteric or hydrothermal mineral and is not formed by weathering.

Glauconite

$KMg(Fe,Al)(SiO_3)_6.3H_2O$ Monoclinic(?)

$$n_\alpha = 1.590 \text{ to } 1.612$$
$$n_\beta = 1.609 \text{ to } 1.643$$
$$n_\gamma = 1.610 \text{ to } 1.644$$
$$2V = 16 \text{ to } 30°; \text{ Opt. } (-)$$
$$\alpha \text{ or } X \ ca. \perp \{001\}$$

Color.—Green, yellow-green, or olive-green in thin sections. Pleochroic from yellow to green.

Form.—Glauconite occurs in grains or pellets that are in part aggregates of minute crystals and in part single crystals. Euhedral crystals have not been observed. The grains are often casts of foraminiferal tests.

Cleavage perfect in one direction $\{001\}$.

Relief moderate, $n >$ balsam.

Birefringence moderate to rather strong, $n_\gamma - n_\alpha = 0.020$ to 0.032; the highest interference colors are second-order colors, but they are masked by the color of the mineral. Many specimens show aggregate polarization.

FIG. 357.—($\times$50) Glauconite (dark) in arkose.

Extinction.—The extinction with reference to cleavage traces is practically parallel, but angles of 2 or 3° have been recorded.

Orientation.—The cleavage traces are length-slow as in the micas.

Interference Figure.—Cleavage flakes give a biaxial negative figure with a small axial angle, but the figure is difficult to obtain on account of the small size of the crystals. Dispersion, $r > v$.

Distinguishing Features.—Glauconite much resembles some of the leptochlorites such as chamosite, but the latter has a higher index of refraction and weaker birefringence. Chamosite usually has an oolitic structure lacking in glauconite.

Alteration.—Glauconite is sometimes altered to limonite.

Occurrence.—Glauconite occurs in sands, sandstones, and limestones. It is especially abundant in the loosely consolidated sandstone known as *greensand*, which is prominent in the Cretaceous of New Jersey. A common associate is collophane.

Origin.—Glauconite is the product of interstitial sedimentation. In the opinion of E. W. Galliher, practically all glauconite is the result of the alteration of detrital biotite.

THE ZEOLITES

The zeolites are hydrous sodium calcium aluminum silicates that commonly occur as secondary minerals in cavities of sub-

ZEOLITES

Mineral	Chemical composition	Crystal system	$n\alpha$	$n\beta$	$n\gamma$
Analcime	Na	Isometric		$n = 1.487$	
Heulandite	Ca	Monoclinic	1.496–1.499	1.497–1.501	1.501–1.505
Stilbite	Ca,Na	Monoclinic	1.494–1.500	1.498–1.504	1.500–1.508
Chabazite	Ca,Na	Monoclinic	$n\alpha = 1.478$–1.485, $n\gamma = 1.480$–1.490		
Natrolite	Na	Orthorhombic	1.473–1.480	1.476–1.482	1.485–1.493
Mesolite	Na,Ca	Monoclinic	1.505	1.505	1.506
Thomsonite	Na,Ca	Orthorhombic	1.512–1.530	1.513–1.532	1.518–1.542
Scolecite	Ca	Monoclinic	1.512	1.519	1.519

silicic volcanic rocks, especially basalts. Although variable in optical properties, they all have low indices of refraction and rather weak birefringence. The last four in the list are fibrous or columnar, but there are also other fibrous zeolites such as ptilolite, mordenite, and laumontite. A few rare zeolites such as harmotome and brewsterite contain barium.

Analcime

NaAl(SiO$_3$)$_2$.H$_2$O   (Analcite)   Isometric

$$n = 1.487$$

Color.—Colorless in thin sections.

Form.—Analcime occurs in equant crystals of trapezohedral habit that are octagonal to rounded in sections. It may also occur in the groundmass in irregular masses.

Cleavage imperfect cubic, which in sections often appears as two sets of lines at right angles.

Relief moderate, $n <$ balsam.

Birefringence.—Analcime is either dark between crossed nicols or shows very weak birefringence (not over 0.002). Use the sensitive-violet plate to detect the double refraction.

Distinguishing Features.—Leucite very much resembles analcime but has a slightly greater refractive index (1.508 as against 1.487).

Occurrence.—Analcime is a secondary mineral in cavities and seams of igneous rocks, usually associated with other zeolites and calcite. In some igneous rocks, such as teschenites and analcime basalts, it occurs in the groundmass as a deuteric mineral. In several western localities it occurs in lake beds.

Heulandite

H$_4$CaAl$_2$(SiO$_3$)$_6$.3H$_2$O   Monoclinic
$$\angle\beta = 88°34'$$

$$n_\alpha = 1.496 \text{ to } 1.499$$
$$n_\beta = 1.497 \text{ to } 1.501$$
$$n_\gamma = 1.501 \text{ to } 1.505$$
$$2V = 0 \text{ to } 48°; \text{ Opt. } (+)$$
$$b = \gamma \text{ or } Z, c \wedge \beta \text{ or } Y = -6°$$

Color.—Colorless in thin sections.

Form.—Heulandite usually occurs in distinct crystals that are tabular parallel to {010}.

Cleavage perfect in one direction {010}.

Relief rather low, $n <$ balsam.

Birefringence weak, $n_\gamma - n_\alpha = 0.007$; the interference colors range up to white of the first order. Sections parallel to {010} that include cleavage flakes have very weak birefringence since $n_\beta - n_\alpha = 0.001$.

Extinction parallel to the cleavage traces.

Orientation.—Cleavage traces are parallel to the faster ray.

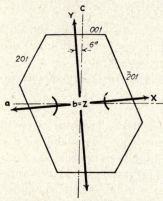

Fig. 358.—Orientation diagram of heulandite. Section parallel to (010).

Interference Figure.—The figure is biaxial positive with a moderate axial angle. The axial plane is normal to {010}. Dispersion, $r < v$.

Distinguishing Features.—Heulandite resembles stilbite but has better cleavage and is optically positive, whereas stilbite is optically negative. The side pinacoid {010} sections of heulandite are unsymmetrical and the corresponding sections of stilbite symmetrical on account of twinning.

Occurrence.—Heulandite is a secondary mineral in the seams and cavities of igneous rocks, especially basalts. Stilbite is a common associate.

Stilbite

$H_4(Ca,Na_2)Al_2(SiO_3)_6 \cdot 4H_2O$

Monoclinic
$\angle\beta = 51°$

$$n_\alpha = 1.494 \text{ to } 1.500$$
$$n_\beta = 1.498 \text{ to } 1.504$$
$$n_\gamma = 1.500 \text{ to } 1.508$$
$$2V = 33° \pm; \text{ Opt. } (-)$$
$$b = \beta \text{ or Y, } a \wedge \alpha \text{ or X} = +5°$$

Color.—Colorless in thin sections.

Form.—Stilbite usually occurs in sheaf-like aggregates.

Cleavage good in one direction {010}.

Relief rather low, $n <$ balsam.

Birefringence weak, $n_\gamma - n_\alpha = 0.006$ to 0.008; interference colors are gray and white of the first order.

Extinction.—Extinction of sections showing the best cleavage is parallel. The extinction angle of sections with the highest interference colors is about 5°. The extinction is usually wavy and not uniform.

Orientation.—The cleavage traces are parallel either to the slow ray or to the fast ray.

Twinning.—Twins with {001} as twin-plane are common.

Interference Figure.—The figure is biaxial negative with a moderate axial angle. The axial plane is {010}. Dispersion, $r < v$.

Distinguishing Features.—Heulandite is similar to stilbite, but it has better cleavage and is optically positive instead of negative.

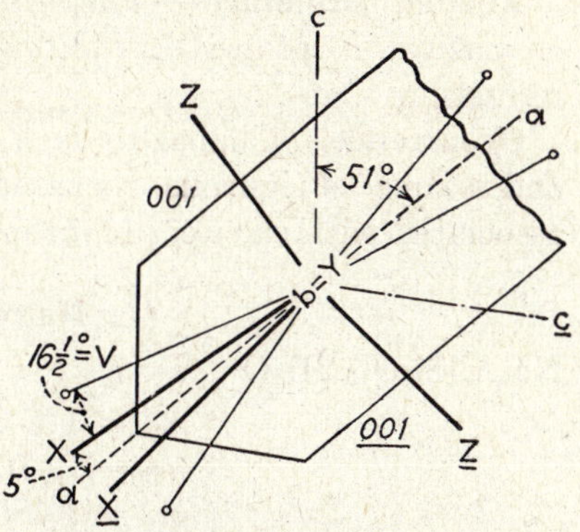

FIG. 359.—Orientation diagram of twinned stilbite. Section parallel to (010); twin-plane = (001).

Occurrence.—Stilbite is a secondary mineral in cavities and seams of igneous rocks. Usual associates are calcite, heulandite, and other zeolites. It has been found as a hot-spring mineral in the interstices of sandstone.

Chabazite

$(Ca,Na_2)Al_2(SiO_3)_6.6H_2O$

Monoclinic
(Pseudo-rhombohedral)

$$n_\alpha = 1.478 \text{ to } 1.485$$
$$n_\gamma = 1.480 \text{ to } 1.490$$
$$2V = 0 \text{ to } 32°; \text{ Opt. } (+)$$

Color.—Colorless in thin sections.

Form.—Chabazite is usually found in euhedral rhombohedral crystals that approach the cube ($10\bar{1}1:\bar{1}101 = 85°14'$).

Cleavage imperfect rhombohedral, hence almost rectangular.

Relief moderate, $n <$ balsam.

Birefringence very weak to weak, $n_\gamma - n_\alpha = 0.002 \text{ to } 0.010$; interference colors are first-order gray.

Extinction.—The extinction is symmetrical to crystal outlines and cleavage traces.

Interference Figure.—The figure is either uniaxial or biaxial with a small axial angle. The optical character is positive.

Distinguishing Features.—Chabazite may be mistaken for other zeolites, especially analcime. The birefringence of chabazite is a little higher than that of analcime.

Related Minerals.—Gmelinite is a zeolite very similar to chabazite in properties, but with slightly lower indices of refraction.

Occurrence.—Chabazite is a secondary mineral in cavities and seams of igneous rocks, especially basalts. It is often associated with calcite, prehnite, and other zeolites.

Natrolite

$Na_2Al_2Si_3O_{10}.2H_2O$ Orthorhombic

$$n_\alpha = 1.473 \text{ to } 1.480$$
$$n_\beta = 1.476 \text{ to } 1.482$$
$$n_\gamma = 1.485 \text{ to } 1.493$$
$$2V = 60 \text{ to } 63°; \text{ Opt. } (+)$$
$$a = \alpha \text{ or } X, b = \beta \text{ or } Y, c = \gamma \text{ or } Z$$

Color.—Colorless in thin sections.

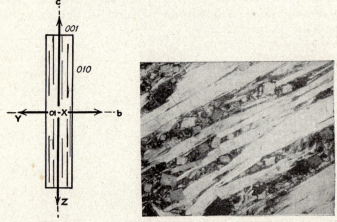

Fig. 360.   Fig. 361.
Fig. 360.—Orientation diagram of natrolite. Section parallel to (100).
Fig. 361.—(×12) Natrolite in rock section.

Form.—Natrolite usually occurs in long prismatic crystals or fibrous aggregates that are often more or less radiating. Cross sections of crystals are nearly square (110 $\wedge$ 1$\bar{1}$0 = *ca.* 89°).

Cleavage parallel to the length of the crystals {110}.

Relief moderate, $n <$ balsam.

Birefringence rather weak, $n_\gamma - n_\alpha = 0.012$ to 0.013; the maximum interference color is yellow or orange of the first order.

Extinction parallel in longitudinal sections, symmetrical in cross sections.

Orientation.—The crystals are always length-slow.

Interference Figure.—A good figure is difficult to obtain on account of the small size of most of the crystals.

Distinguishing Features.—Scolecite resembles natrolite but is length-fast instead of length-slow and has oblique extinction. Thomsonite has parallel extinction but is length-slow in some sections and length-fast in others.

Occurrence.—Natrolite is a secondary mineral found in cavities of igneous rocks, especially basalt. The associates are other zeolites and calcite.

Mesolite

$Na_2Ca_2Al_6(Si_3O_{10})_3.8H_2O$ Monoclinic

$$n_\alpha = 1.505$$
$$n_\beta = 1.505$$
$$n_\gamma = 1.506$$
$$2V = ca.\ 80°;\ Opt.\ (+)$$
$$c \wedge \beta \text{ or } Y = 2 \text{ to } 5°$$

Color.—Colorless in thin sections.

Form.—Mesolite usually occurs in fibrous aggregates.

Cleavage perfect in two directions $\{110\}$ and $\{1\bar{1}0\}$.

Relief moderate, $n <$ balsam.

Birefringence very weak, $n_\gamma - n_\alpha = 0.001$; the maximum interference color is first-order gray.

Extinction.—The maximum extinction angle in longitudinal sections is very small, from 2 to $5°$.

Orientation.—The fibers are in part length-slow and in part length-fast.

Twinning.—Twins with $\{100\}$ as twin-plane are universal, but the twinning is not conspicuous.

Interference Figure.—The figure is biaxial positive with a very large axial angle. The figure lies across the fibers. Dispersion, $r > v$ strong.

Distinguishing Features.—Mesolite very much resembles the other fibrous zeolites. In common with thomsonite the fibers are in part length-slow and in part length-fast. From thomsonite it may be distinguished by the maximum extinction angle of 2 to 5° and by its larger axial angle.

Occurrence.—The occurrence of mesolite is the same as that of other zeolites, in the cavities of basalts and related rocks.

Thomsonite

$NaCa_2Al_5(SiO_4)_5.6H_2O$ Orthorhombic

$$n_\alpha = 1.512 \text{ to } 1.530$$
$$n_\beta = 1.513 \text{ to } 1.532$$
$$n_\gamma = 1.518 \text{ to } 1.542$$
$$2V = 44 \text{ to } 55°; \text{ Opt. } (+)$$
$$a = \alpha \text{ or } X, b = \gamma \text{ or } Z, c = \beta \text{ or } Y$$

Color.—Colorless in thin sections.

Form.—Thomsonite usually occurs in fibrous or columnar aggregates. Euhedral crystals are very rare.

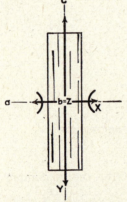

    Cleavage in one direction {010}.

    Relief rather low, $n <$ balsam.

    Birefringence rather weak, $n_\gamma - n_\alpha = 0.006$ to 0.012; maximum interference colors range from first-order white up to low second-order blue in different specimens. Cross sections of fibers show the highest interference color for a given thickness.

    Extinction parallel.

    Orientation.—Some of the fibers are length-slow and some length-fast since

Fig. 362.—Orientation diagram of thomsonite. Section parallel to (010). $c = Y$.

    Interference Figure.—The figure is biaxial positive with rather large axial angle. The figure lies across the fibers since the axial plane is {001}. Dispersion, $r > v$ strong.

Distinguishing Features.—Thomsonite is much like the other fibrous zeolites in general appearance and optical properties. Natrolite is length-slow and scolecite length-fast, whereas some of the fibers of thomsonite are length-slow and some length-fast. The same is true of mesolite, but in mesolite the maximum extinction angle $c \wedge \beta$ or Y is about 3°. The axial angle of mesolite is much larger than that of thomsonite.

Occurrence.—Thomsonite occurs as a cavity filling in subsilicic volcanic rocks such as amygdaloidal basalts.

Scolecite

$CaAl_2Si_3O_{10}.3H_2O$ Monoclinic
$\angle \beta = 89°18'$

$$n_\alpha = 1.512$$
$$n_\beta = 1.519$$
$$n_\gamma = 1.519$$
$$2V = 36°; \text{ Opt. } (-)$$
$$b = \gamma \text{ or } Z, c \wedge \alpha \text{ or } X = -15 \text{ to } -18°$$

Color.—Colorless in thin sections.
Form.—Scolecite occurs in crystal aggregates with a columnar to fibrous structure.

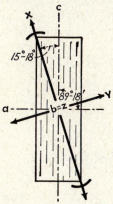

Fig. 363.—Orientation diagram of scolecite. Section parallel to (010).

Fig. 364.—(×22) Scolecite.

Cleavage distinct in two directions {110} at angles of *ca.* 88°.
Relief low, $n <$ balsam.
Birefringence weak, $n_\gamma - n_\alpha = 0.007$; so the interference colors are gray and white of the first order.

Extinction.—The maximum extinction angle in longitudinal sections is -15 to $-18°$.

Orientation.—The crystals are always length-fast.

Twinning.—Twinning is common. The c-axis [001] is the twin-axis and (100) the composition face.

Interference Figure.—The figure is biaxial negative with a moderate axial angle. The axial plane is normal to {010}. Dispersion, $r < v$ strong.

Distinguishing Features.—Scolecite is much like natrolite and other rare fibrous zeolites. The oblique extinction of about 17° and twinning are the most distinctive features of scolecite.

Occurrence.—Scolecite is a secondary mineral found in cavities of igneous rocks, especially basalts.

THE LESS DEFINITE MINERALOIDS

The term *mineraloid* is given to the mineral-like constituents of rocks that are not definite enough in chemical composition or in physical properties to be included in the list of minerals. The most common and abundant mineraloid is naturally occurring glass or volcanic glass, which is widespread and often of geological importance. Obsidian, perlite, pitchstone, pumice, etc., are petrographic terms, but glass as a whole may be treated as a mineraloid.

Palagonite, an alteration product of fragmental basaltic glass formerly classed as a mineral, is undoubtedly a mineraloid.

Hydrocarbons also may be treated as mineraloids. Hydrocarbons are very difficult to obtain in thin sections on account of their solubility in balsam.

VOLCANIC GLASS

$SiO_2, Al_2O_3, Fe_2O_3, FeO, MgO,$ Amorphous
$CaO, Na_2O, K_2O, H_2O,$ etc.   (Mineraloid)

$$n = 1.48 \text{ to } 1.61$$

Color.—Colorless to gray or reddish in thin sections.

Form.—Usually massive, sometimes vesicular, perlitic, etc. Often contains spherulites of orthoclase, microlites, crystallites, microphenocrysts, and phenocrysts.

Cleavage absent, but it may show perlitic parting.

Relief low to moderate, n usually less than balsam but sometimes greater. The index of refraction increases as the silica decreases.

Birefringence usually nil, but some varieties show weak birefringence that is due to strain.

Fig. 365.—(×12) Volcanic glass (pumice) showing flow texture.

Fig. 366.—(×12) Volcanic glass (perlite) showing perlitic texture and crystallites.

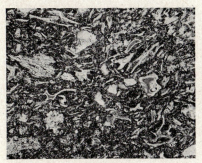

Fig. 367.—(×30) Glass shards in rhyolite tuff.

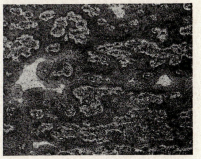

Fig. 368.—(×54) Devitrified glass (obsidian) with imperfect spherulites and relicts of glass.

Distinguishing Features.—Opal may be mistaken for glass, but the refractive index of opal is distinctly lower.

Alteration.—Volcanic glass is often more or less devitrified (see Fig. 368, above). The alteration products are usually rather indefinite, but sometimes feldspars, tridymite, cristobalite, or montmorillonite are the result of devitrification. Palagonite is always the result of alteration of glass fragments.

Occurrence.—Glass often occurs as an independent igneous rock such as obsidian, pumice, perlite, or pitchstone. Most volcanic glass corresponds to rhyolite in composition. Glass is also found as a narrow selvage to basalt dikes. This variety is known as *tachylyte*. Glass is a prominent constituent of vitrophyre and occurs in the groundmass of many volcanic rocks.

Silica glass has been described under the name *lechatelierite* (see page 192).

Palagonite

$SiO_2, Al_2O_3, Fe_2O_3, FeO, MgO,$ Amorphous
CaO, H_2O (Mineraloid)
(Altered Glass)

$$n = 1.47 \text{ to } 1.63$$

Color.—Usually yellow to yellowish brown but also brown and greenish in thin sections.

Form.—Palagonite is found as a rim or zone around glass fragments or in massive form. It often shows an apparent oolitic structure that is due to the filling of microvesicles.

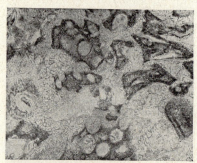

Fig. 369.—(×56) Palagonite (dark) showing microvesicles with aragonite in tuff.

Relief low to medium, n either less or greater than balsam.

Birefringence nil to very weak. Palagonite is a hydrogel, and the weak birefringence sometimes noted is probably due to strain.

Distinguishing Features.—Palagonite resembles opal, collophane, and volcanic glass. The index of refraction is too high for opal and nearly always too low for collophane. It is distinguished from glass by the high water content.

Alteration.—The palagonite formed from basaltic breccia fragments at Roseburg, Ore., has, according to A. C. Waters, been altered to chlorite.

Occurrence.—Palagonite is found in palagonite tuffs and palagonite rock. It is formed by the hydration of fragmental basaltic glass. The glass fragments are in whole or in part converted into the palagonite. Associates besides glass are calcite, zeolites, chlorite, pyroxene, olivine, and plagioclase (the last three are relict minerals from the original glass).

INDEX

A

Abbe refractometer, 49–51
Abbreviations, xv, xvi
Abrasives, 4, 5
Absorption, 114
  formula, 114
Accessories, microscope, 16–19
Achromatic objective, 14
Acmite, 271
ACTINOLITE, 282
Acute bisectrix, 96
  figures, 95, 97
Adjustment, fine, 18
  of crosshairs, 29
  of microscope, 28–31
  of nicols, 29
Adularia, 233
Aegirine, 270
Aegirine-augite, 269
Aegirite (=aegirine), 270
Aggregate structure, 117
Aggregates, fine, 117–119
  fibrous, 119
Aggregation, 114–121
Air, index of, 42
Åkermanite, 257, 258
ALBITE, 246
  twin law, 239
Allanite, 326
Alleghanyite, 301
Allen, R. M., 31
Allophane, 352, 353, 354
Almandite, 301, 302
Alnöite, 258, 299
Alundum, 5
Alunite, 221
Ammoniojarosite, 222
Amorphous minerals, 115
Amphibole group, table of, 277

AMPHIBOLES, 276–291
  cleavage of, 128
  crystal form of, 126
  twinning in, 121
Amplitude of light waves, 35
Analcime, 367
  crystal form of, 123
Analcite (=analcime), 367
Analyzer, 16
Anatase, 198
Anauxite, 352, 353, 354
Andalusite, 308
  crystal form of, 125
ANDESINE, 248
Andradite, 301, 302
Angle, axial, 104–106
  critical, 44–45
  of extinction, 79–82
  of incidence, 41, 42, 44
  of minimum deviation, 52
  of refraction, 42, 45
  of rhombic section, 239, 245
Anhedral form, 123
Anhydrite, 216
Anisotropic minerals, 46
  index of refraction of, 46
  tables of, 164–169
Anisotropism, due to strain, 82
Ankerite, 209
Anomalous interference, 82–83
Anomite, 342
Anorthite, 250
Anorthoclase, 237
Anthophyllite, 277
  hydrous, 278
ANTIGORITE, 361
APATITE, 224
  crystal form of, 121, 125
Apochromatic objective, 14
Aragonite, 213
  axial angle of, 106

377

Canada balsam, 5
Cancrinite, 254
Carbonaceous matter, 179
Carbonates, rhombohedral, 206–212
Carborundum, 5
Care of microscope, 21
Carlsbad twin law, 121, 239
Cassiterite, 198
  twinning of, 121
Celestite, 215
Cellular structure, 117
Celsian, 230
Centering of microscope stage, 28
Centering pin, 16, 17
Central illumination, method of, 55
Chabazite, 369
CHALCEDONY, 188
  structure, mosaic, 117, 118
    radial, 117, 118
Chalcopyrite, 182
Chamosite, 347
Chamot, E. M., 31
Charnockite, 262
Chiastolite, 308
CHLORITE GROUP, 342
Chloritoid, 348
Chloromelanite, 272
Chondrodite, 299
Chromite, 201
Chrysolite (=Olivine), 294
Chrysotile, 363
  asbestos (Fig.), 120
Chudoba, K., 176
Circular sections of ellipsoid, 102
Classification of minerals, 177–178
Clay films, 7
CLAY MINERALS, 352–360
Cleavage, 80, 126–130
  cubic, 128
  dodecahedral, 130
  octahedral, 129
  rectangular, 129
  relation to form, 140
  table of, 155–156
Cliachite, 204
  amorphous character, 115
  pisolitic form, 115
Clinochlore, 343

Clinoenstatite, 267
Clinohumite, 301
CLINOPYROXENES, 262–273
Clinozoisite, 321
Coker, E. G., 39
Colloform structure, 117
COLLOPHANE, 226
  amorphous character, 115
  bone structure in, 117
Color, effect of vibration axes on, 114
  of light, 38
  of minerals, 113–114, 175
  neutral, 175
Color bands, of interference figures, 89–90
Color chart, *facing* 163
Compensator, Berek, 19, 75–78
Condenser, 18, 85
Condenser circle, 28
Conoscopic study, 10
Convergent polarized light, 85–112
Cookeite, 274
Cordierite, 330
Corpuscular theory of light, 32
Corundum, 194
  skeleton crystals of, 121
  twinning in, 121
Cover glass, 20
Crest of light wave, 35
Crew, H., 39
Cristobalite, 191
  artificial, 192
Critical angle, 44–45, 48
Crocidolite, 289
Crookes, W., 82
Crosshairs, adjustment of, 29–30
Crossite, 276, 291
Crushed fragments, 135
Crystal form, in thin section, 121–126
Crystal systems, 121–126
Crystallites, 115
Crystallization, incipient, 115
Crystals, anhedral, 123
  bladed, 121
  euhedral, 123
  hexagonal, 123